Robert Jones, John Ridlon

Contributions to Orthopedic Surgery

Robert Jones, John Ridlon

Contributions to Orthopedic Surgery

ISBN/EAN: 9783337089832

Printed in Europe, USA, Canada, Australia, Japan

Cover: Foto ©berggeist007 / pixelio.de

More available books at **www.hansebooks.com**

TO

ORTHOPEDIC SURGERY

BY

ROBERT JONES, F.R.C.S.E.

Hon. Surgeon to the Royal Southern Hospital. Liverpool :
Hon. Surgeon to the Liverpool Country Hospital for Children :

AND

JOHN RIDLON, M.A. M.D.

Professor of Orthopedic Surgery in the North-Western University Medical Schools :
Senior Orthopedic Surgeon St. Luke's and Michael Reese Hospitals,
and the Home for Crippled Children. Chicago :
Secretary American Orthopedic Association.

THIS WORK

IS DEDICATED WITH INFINITE LOVE AND RESPECT
TO THE MEMORY OF

HUGH OWEN THOMAS

BY THE AUTHORS.

PREFACE.

AT intervals during the last twelve years we have written many papers on the surgery of joints and deformities. Urged by our friends, we now publish some of these in a more permanent form. The book is merely intended for house surgeons and others who, knowing our methods, desire some guidance as to the principles which underlie them. We have therefore, for the time being, contented ourselves by reprinting old essays, having no intention to submit them to the profession generally or to the reviewer's pen. Before long it is intended to present an amended and more complete work.

ROBERT JONES.

JOHN RIDLON.

TABLE OF CONTENTS.

GENERAL PRINCIPLES RELATING TO CHRONIC JOINT-DISEASE.

Chronic joint-disease is the term generally used to indicate a chronic pathologic process affecting any of the structures of a joint, which, untreated, may be expected to result in permanent disability or in deformity. The structure first involved is usually the bone; next in order of frequency is the synovial membrane; but the source whence the pathologic condition arises generally determines the first structure to be involved.

The pathologic process in chronic joint-disease is usually a primary tubercular inflammation, or a secondary tubercular inflammation transmitted from some primary focus located elsewhere, such as a cheesy bronchial lymph-node, or it may be a tubercular inflammation engrafted upon a local traumatic or syphilitic inflammation. It is possible that tuberculosis may become engrafted upon a rheumatic inflammation, but from clinical experience this appears to be doubtful.

A tubercular inflammation is the reaction which accompanies local tubercular infection. It is characterized by the growth and development of miliary tubercles. Its progress is usually slow; after reaching a certain stage it is usually slowly retrogressive.

A traumatic inflammation of a joint is the reaction which takes place after an injury, such as a sprain, a fracture into a joint, or a dislocation. The tendency in healthy individuals always is toward recovery; and in severe injuries, such as dislocations, the temporary disability is so great that perfect resolution of the inflammation takes place during the period of enforced rest. This also happens in cases in which fractures extend into joints, unless an over-anxious or meddlesome surgeon delays the resolution of the inflammation by frequent manipulations of the part, as in the use of passive motion.

Passive motion was at one time generally employed when fracture occurred near or extended into a joint, in the belief that it was necessary to prevent stiffness of the joint. We are, however, very positively of the opinion that nothing is gained by early passive motion, but, on the other hand, that forced flexions and extensions of a joint, while the inflammatory reaction, evidenced by tenderness and muscular spasm, remains, not infrequently result in a continuance of the inflammation, and ultimately in tuberculosis of the joint.

Inherited syphilis plays an important rôle in the etiology of chronic joint-disease. Just what the relations are between inherited syphilis and tuberculosis we do not know. From clinical observation, however, it would appear that the disease may begin as a syphilitic focus and ultimately become infected with tubercle-bacilli; and any focus located at an epiphyseal line may fairly be suspected of being, or of having been, syphilitic. Even when it cannot be demonstrated that syphilis itself has been inherited, the constitutional defect which is often found in the children of syphilitics renders them fit subjects for tubercular infection.

Tubercular joint-disease begins in either the bone or the synovial membrane; there is no evidence to show that it ever begins in the ligaments or in the cartilage. Tuberculosis of the bone in the neighborhood of a joint almost invariably begins as a primary or a secondary focus.

A primary focus in the bone is believed to arise in the following manner: A few tubercle-bacilli carried in

the blood-current gather together at some point where
the blood-current becomes sluggish, or where there is
actual stasis, somewhat after the manner that sticks
and leaves gather in an eddy at the side of the current
in a stream; here the bacilli colonize, and a gray miliary
tubercle results; from this other tubercles develop, and
the focus grows. Usually the development is uniform
on all sides, and a globular focus results, but occasionally it is more or less elongated.

A secondary focus in the bone is believed to result
from the occlusion of a terminal artery by a cheesy
particle coming from some previously existing tubercular lesion. This previously existing tubercular lesion
has more often than otherwise been found to be in the
bronchial lymph-nodes, although occasionally the oldest
lesions are found in the abdominal lymphatics. When
the bronchial lymphatics are involved it is believed that
the infection found entrance by way of the lungs, and
by way of the digestive tract when the abdominal
lymphatics are the parts first diseased. When a terminal artery in the end of a bone becomes occluded by
a tubercular particle, the area supplied by the occluded
vessel becomes a fertile field for the growth of the
tubercle bacillus, and the progress of the disease is from
the infecting particle in this direction. As a result, a
cone-shaped focus is ultimately found, with its base
resting against the cartilage of the joint and its apex at
the point where the artery was plugged. This is also
called a triangular focus, from its triangular surface on
vertical section, and it is also called an infarction focus.

The development of the focus, except as to its shape,
is the same in either the primary or the secondary form.
The surface of the section of a focus is grayish-red, yellowish-white, or yellow, the boundary being somewhat
reddened by collateral hyperemia. In the very early
stage the transparent gray tubercles may, with the aid

of a lens, be séen throughout the spot; later on they will only be found at the periphery, the central part having already become cheesy. This caseation appears to result from the extending occlusion of the blood vessels by the pressure of the growing tubercles. Thus death first appears in the center of the mass, and, as the growth advances at the surface, the central area of necrobiosis also increases, and is still found to be surrounded by a gray, or grayish-violet, membrane, easily separable from the surrounding tissues. In this outer membrane bacilli may be found in abundance, while in the cheesy mass filling the cavity they can rarely be demonstrated by microscopic examination or by cultivation; nevertheless, inoculation of a rabbit's cornea with this central cheesy material invariably gives rise to tubercle, forcing us to the conclusion that spores exist therein.

As a rule, the caseating process extends, and continues to undergo liquefaction centrally, until it reaches the surface of the bone, where it finds exit through the periosteum, or through the cartilage and synovial membrane into the joint. Occasionally drying or dehydration takes place, and the spot may even calcify and remain unchanged for years, the surrounding hyperemia leading to exudation of round cells and the formation of a fibrous capsule, or to sclerosis of the bone. In other cases resolution may take place before the cheesy stage has been reached; or cicatricial healing may result after suppuration or caseation by the development and encroachment of healthy granulation-tissue; sequestra of very considerable size may disappear, and only the scar remain, after the manner of an ordinary infarct in the spleen or the kidney.

Tuberculosis of the synovial membrane is known to occur as secondary to adjacent bone-tuberculosis; and it is also believed to occur primarily by infection from the blood.

Primary synovial infection is believed to occur in all cases when the synovial disease has a traumatic etiological factor; but it is also seen clinically when there is no history of traumatism and no history of an osseous lesion. In these cases the older and yellow tubercles are found on the surface of the membrane, and the more recent gray tubercles deeper within the substance of the membrane.

Secondary synovial infection may result, as already stated, from the rupture of a liquefied osseous focus, through the cartilage and synovial membrane. Under such circumstances, the older yellow tubercles are found on the surface of the membrane, and the more recent gray tubercles deeper within its substance.

A local secondary infection of the synovial membrane may also result from the extension of the osseous focus without the rupture of the liquefied cheesy mass. Then the older tubercles are found deeply imbedded, and the more recent ones are at or near the surface.

A third and more rare form of joint-tuberculosis is believed to commence as a primary synovial tuberculosis and to proceed rapidly to destruction of the entire articular cartilage and infiltration of the exposed spongy portion of the bone; it may make its way through the entire diaphysis and into the medullary cylinder itself.

The liquefied cheesy focus is called a tubercular abscess. When it breaks through the periosteum, it spreads beneath the soft parts, and in the case of superficial joints may soon be made out by palpation. When it ruptures into a joint the disease may go on to the rupture of the synovial sac or not. If the synovial sac eventually ruptures we have the same condition of affairs as exists when a tubercular abscess has made its way through the periosteum.

The tubercular abscess, when it has escaped from the

bone or from the joint, develops comparatively slowly. Many months may elapse before it finds its way to the surface; it follows the direction of least resistance, influenced by gravity; the size of the abscess bears no direct relation to the bony focus; an abscess containing one or two quarts may have had its start from a bone-focus the size of a pea, and the bone-focus may have healed long before the abscess finds its way to the surface. These abscesses present neither heat, redness (except when about to break), tenderness, nor pain.

The fluid contents appear differently at different stages. Early it is thick, creamy, and yellow; but it lacks the greenish color and viscidity of phlegmonous pus. Later on we find minute yellow or yellowish-white flocculi floating in a thinner turbid fluid; and still later curds or clots of fibrin and shreds of connective tissue are found floating in a more or less clear, whey-colored liquid. The tubercular pus contained in these abscesses presents no microorganisms on staining or cultivation, but inoculation of the pus in fit subjects results in the development of tuberculosis.

The walls of these abscesses are made up of grayish-yellow or violet membrane (the old "pyogenic membrane") of greater or less thickness, always easily detachable, consisting of a soft brittle tissue composed essentially of closely aggregated miliary tubercles imbedded in fibrin, the cavity itself being lined with this material. The sinuses formed by the spontaneous opening of these abscesses, and those which form after incision, are lined with the same membrane, its thickness appearing to depend very much upon the amount and duration of the irritation to which the abscess-cavities and sinuses have been subjected.

The predisposing causes of tubercular joint-disease are not limited solely to the local conditions, and to inherited syphilis and to traumatism, already mentioned. Congenital tuberculosis cannot be denied, although undoubtedly the condition is a rare one. On the other hand a constitutional predisposition is, as a rule, handed down from the tuberculous parent to the child ; and the same constitutional predisposition is at times found in several of the children of a family having non-tuberculous parents and grandparents. Tuberculosis of the joints may occur after the infectious diseases of childhood, such as measles, whooping-cough, and scarlet fever, and after exhausting diseases, parturition, privation, and prolonged dissipation ; but the best conditions of nutrition offer no certain protection against its occurrence.

The general symptoms of tubercular joint-disease, speaking broadly, fall into one or the other of three classes :

1. Primary synovial disease of non-traumatic origin.

2. Bone-disease either from primary infection or secondary to disease in other tissues, associated or not with traumatism.

3. A purely traumatic lesion, such as a sprain,which, in the course of time, has become chronic from lack of treatment.

The joint of the first class presents, at least for a time, only a distended joint-capsule, the normal bony outlines being indistinct or completely lost, with true or false fluctuation on palpation. There is to the touch no local elevation of temperature, no limitation of motion except the mechanical limitation due to the thickened and distended capsule, no muscular atrophy, no limping except after prolonged use, no complaint of pain, and usually no tenderness.

The joint of the second class presents no distended capsule, the bony outlines being normal ; there often is a local elevation of temperature, and a tender point can be made out when a superficial joint is involved ;

limping begins early, and is a constant symptom when
a joint of the lower extremities is involved, and disability is present when a joint of the upper extremities
or the spine is affected, although intermissions in these
symptoms are occasionally met with; muscular shrinking and more or less restriction of joint-motion are
always found; pain comes on, after a time, in most
cases, but may be absent throughout the entire course
of the disease.

The joint of the third class is characterized by the
symptoms peculiar to traumatism. Usually .it is a
sprain, for rarely does a fracture into a joint, or a dislocation, result in chronic disease. There is swelling
and infiltration of the soft parts about the joint, with
either true or false fluctuation; increased heat can
always be felt; local pain of an aching character is
almost invariably present; muscular atrophy and
restriction of motion are always found; and, finally,
there is a general tenderness to pressure rather than a
small and sharply defined tender point.

As the disease progresses in either class, the three
pictures above shown merge into one another to form
a composite, presenting all the symptoms above enumerated. In addition, there is deformity varying with
the location of the disease. If the spine is affected,
there is a posterior angle or curvature, with the opening
or concavity to the front, and early in the disease there
is a lateral bending in many cases; if the hip, it is
flexed and usually abducted or adducted; if the knee,
it is flexed, and later on the leg is abducted and rotated
outward; if the ankle, there is plantar flexion; if the
shoulder, there is adduction; and if the elbow or wrist,
there is flexion.

Complications: Abscesses appear in nearly half of
all cases, and partial or complete subluxation at the
hip and knee occurs in a few cases. Lardaceous dis-

ease of the liver and kidneys and meningeal tuberculosis are the fatal complications.

The prognosis in tubercular joint-disease must be
considered as to life, as to the duration of the disease,
and as to the ultimate functional result.

As to life: From 8% to 33% die as the result of the
tubercular disease, the variation depending upon the
part involved, the age of the patient, the constitutional
predisposition, and upon the general hygienic surroundings and the orthopedic treatment available.

As to the duration of the disease and the ultimate
functional result: In general, other things being equal,
the younger the patient the shorter will be the duration
and the better will be the ultimate functional result.
Smaller joints recover more quickly and more perfectly
than larger ones. A certain number of joints recover
either under mechanical treatment or under operative
treatment, or without any treatment whatever, with
limbs in fair position and joints possessed of a good
range of motion; and suppuration, whether treated or
untreated, is no bar to this result. On the other hand,
certain cases, no matter how early treatment be commenced, or how carefully carried out, will go on to
recovery with short limbs, or stiff joints, or ultimately
to death. The duration of disease in individual joints
or particular patients cannot be accurately foretold.
Relapses rarely occur if the surgeon recognizes the
signs of perfect recovery. The cases that relapse after
mechanical treatment or operative treatment are those
in which treatment has been suspended before the
articulation has regained perfect soundness.

The treatment of tubercular joint-disease may be
divided into hygienic, medicinal, mechanical, and operative. In the hygienic treatment sunshine, pure air,
bathing, and an abundance of proper food are of chief
importance, and the value of digestible fats should

never be lost sight of. Voluntary physical exercise is not essential to good health, as some have assumed; as a rule, exercise cannot be indulged in without serious risk to the diseased joint in the early treatment of very many cases. The medicinal treatment, both general and local, is far from satisfactory. It has not been demonstrated that any medicine has a direct remedial action. General tonics may be indicated from time to time, but obviously cannot be continued without intermission throughout the months and years that these patients must remain under treatment. So long as the appetite and digestion remain reasonably good we are accustomed to give little or no medicine, except cod-liver oil, unless there be a reasonable suspicion of inherited syphilis. The history of syphilis in the parents can rarely be elicited. One may, however, be warranted in suspecting syphilis in the following cases: All cases of chronic joint-disease in children under three years of age, except when there is a positive history of advanced tuberculosis in the mother at the time the child was born; in all patients suffering from multiple joint-disease; in colored children, and in the children of other classes that are notoriously syphilitic. In such cases we are accustomed to give mercury and potassium iodid in full doses. When the bichlorid or biniodid in solution is not well borne, we are accustomed to give the officinal powder of mercury with chalk, or an inunction of blue ointment may be used. The dose, in any instance, may be as large as would be given to an adult. Potassium iodid may be given in increasing doses, from gr. v to ℨ iij *t. i. d.*

Local external medication, such as liniments, salves, tincture of iodin, blisters, and the cautery, in strictly tuberculous joints, is absolutely useless and at times harmful. Deep injections into the joints, into the tubercular foci, and into the neighborhood of the diseased tissues are too painful to be tolerated by children, and too uncertain in their results for routine treatment. Iodoform suspended in oil or glycerin has attained a certain popularity among surgeons having neither skill nor experience in the use of orthopedic appliances; but there is little evidence to show that the treatment possesses any curative value, while there is abudant evidence showing that in the vast majority of cases it is positively harmful.

Local congestion of the joint and its immediate neighborhood, which has recently come into use in Germany, is, in some cases, of positive advantage. This method of increasing the nutrition locally was first recommended by the late Dr. Hugh Owen Thomas and employed by him for many years. The recent "discoverer" appears to be wholly ignorant of Thomas' writings.

The mechanical treatment of chronic joint-disease aims to protect the diseased joint from injury inflicted by movement at the joint, from shocks during locomotion, from the burden of weight-bearing when the joints of the spine or lower extremities are involved, and from the intra-articular pressure due to the involuntary muscular spasm in Nature's attempts at immobilization.

The first and most important problem is the immobilization of the joint. The materials used are of little importance, provided the essential principles are not lost sight of; namely, to immobilize a joint it is necessary to put at rest the muscles governing motion at that joint, and to do this the immobilizing apparatus must extend to the limits of these muscular attachments; to perfectly protect a joint from the shocks of locomotion and weight-bearing, recumbency must be employed; and to counteract intra-articular pressure resulting from muscular spasm, the irritation inducing the spasm must be allayed. This can be accomplished more often than otherwise by immobilization of the joint, but in some instances traction upon the limb is of very material assistance. In a few instances traction aggravates the pain and muscular spasm, and in the vast majority of cases its sedative action is only as an aid to other forms of immobilization. In the treatment of these joints it may be laid down as a law that whatever most quickly relieves pain and tenderness will most quickly relax the muscular spasm and place the joint in a condition favorable to recovery.

The essential of any orthopedic appliance is that the framework be firm, unyielding, and free from tremor; that the padding be sufficient to protect the soft parts from harmful pressure, nor yet so soft nor so thick as to diminish the effectiveness of the rigid frame; and that the covering be such as will not readily become infected with septic microorganisms.

The operative treatment includes aspiration and incision of abscesses, erasion and excision of joints, and amputation of limbs.

Abscesses should never be subjected to operative interference unless the patient be suffering from septic infection, or the bulk of the abscess prevents effective mechanical support. When from either of these indications it becomes necessary to operate upon a tubercular abscess it should be freely incised, its contents evacuated, its lining membrane rubbed off with a piece of sterile gauze, its bone-focus, if such be found, curetted, the whole cavity washed out and dried, and the wound through the healthy tissues accurately closed throughout its entire extent with numerous sutures. Drainage, and especially drainage by a rubber tube, should not be used. If all or most of the tubercular tissue has been removed, and the operation has been aseptic, primary healing will result; otherwise the abscess will refill and demand a second opening, or open spontaneously, usually through some part of the closed incision. If the wound be still aseptic artificial drainage need not be used ; if it be septic it may be washed out and drained. Prolonged drainage, however, always results in a sinus, with greatly thickened tubercular walls, and which heals only after a very prolonged period.

Erasion consists in opening the joint or bone focus and scraping out the diseased tissue. When the synovial membrane is not involved, and the disease is wholly confined to a bone-focus, erasion is a justifiable procedure, provided the focus can be accurately located. When, however, the synovial membrane is involved, either with or without the presence of a neighboring bony focus, the operation of erasion has not proved satisfactory and is now rarely performed.

Excision of a joint consists in the sawing off and removal of one or both bones going to make up the joint, and the consequent total removal of all diseased tissues in very many cases. The operation can be very readily done at the knee and elbow, and at these joints it succeeds in the majority of cases in eradicating the disease; but at the other joints the results are far less satisfactory. At the hip it is practically impossible to excise the acetabulum and remove all of the disease, and relapses are of frequent occurrence. Among orthopedic surgeons excisions are rarely performed except as a life-saving measure, or as a time-saving measure in adult cases when poverty renders prolonged mechanical treatment impossible. Excision should never be practised in children except as a life-saving measure ; for in children an excision results in an arrest of growth in the limb, sometimes amounting to six or eight inches in cases of excision at the knee.

Amputation is performed only as a life-saving measure. In children it is often preferable to excision.

SPONDYLITIS.

Spondylitis is an inflammation of one or more of the bones of the spine, characterized by stiffness and disability, and sooner or later resulting in a greater or less degree of spinal deformity.

The synonyms in common use are Pott's disease, spinal caries, hump-back, and hunch-back.

Fig. 1.—Caries of lower dorsal spine, showing complete destruction of one of the vertebral bodies. (Krause.)

17

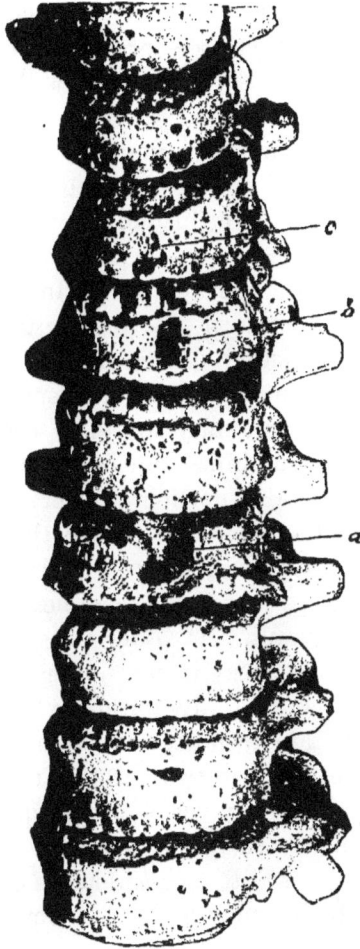

FIG. 2.—Foci of disease starting in anterior surface of vertebral bodies, and separated by one healthy vertebra. (Krause.)

The causes are tuberculosis by infection (common) or by inheritance (rare), inherited syphilis, and injury from falls, blows, and the lifting of heavy weights. The disease also follows and appears to depend upon scarlatina, measles, whooping-cough, and other infectious diseases; but however it begins, or whatever be its specific origin, the symptoms presenting and the indications for treatment are practically almost identical, and all cases sooner or later show the presence of the tubercle-bacillus.

The pathology practically amounts to the deposit of tuberculous material and the subsequent growth and development of the focus, usually in the anterior part of the body of one or the bodies of several vertebræ; very rarely do the symptoms indicate that the tuberculosis commences in the intervertebral discs, or in the laminæ, articular facets, or processes. As the disease develops the spine often presents a bowing or curvature due to involuntary spasm of the muscles in their attempt to immobilize the diseased area. Later on, when a considerable portion of a vertebral body has been softened by the tubercular growth, the bone crushes together and a posterior angle is formed. At times the entire body of a vertebra will disappear, a very acute angle being the result. At other times there will be only a small spot of decay in several vertebræ, when in place of the angular deformity the spine becomes curved posteriorly. In rare instances a well-marked destructive process takes place in two parts of the spine separated by several healthy vertebræ, in which case two angular deformities result. Late in the disease, when healing has progressed to a considerable extent, two or more vertebral bodies may be found consolidated into a confused mass by the deposit of new bone.

The symptoms of spondylitis may be common to the disease in any part of the spine whatsoever or peculiar to the part of the spine affected. The symptoms common to the disease in any part of the spine are as follows: The face expresses apprehension, pain, and premature old age. The patient walks and moves with care, as if to avoid any jar or sudden movement. There can be obtained a history of uneasiness, fretting, and

irritability, and for some time the patient has been dis-
inclined to exercise as actively as usual and has been
easily fatigued. Distant pain, felt in the terminal fila-
ments of the nerves whose motor branches go to supply
the muscles controlling motion of the spine at the point
of disease, has generally been felt, though it may have

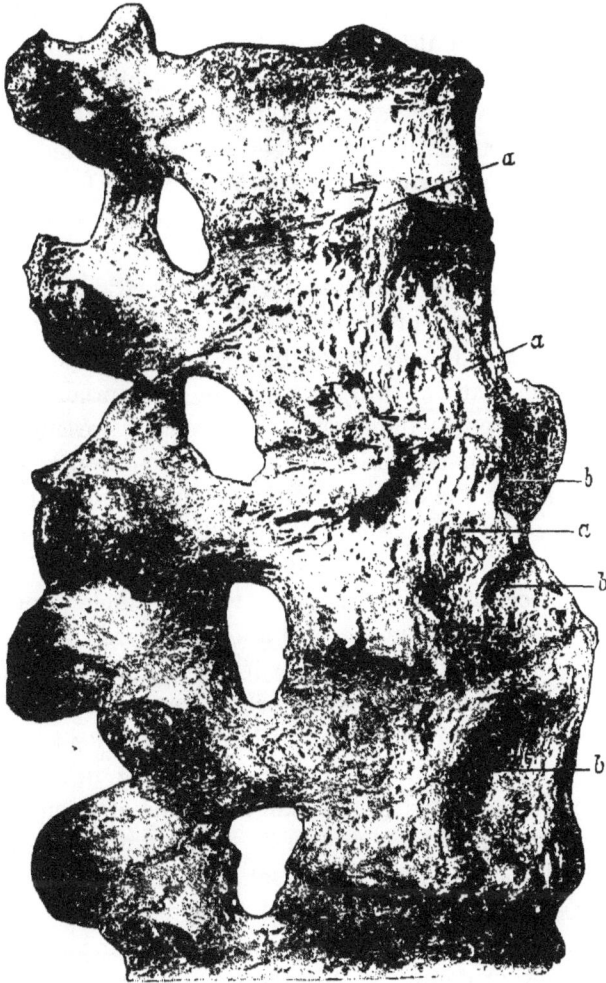

Fig. 3.—Bony deposit and ankylosis in a case of healed spondylitis.
(Krause.)

been absent, as may also have been restlessness, crying and screaming during the first hours of sleep. Deformity may or may not have been noticed, and the complications—abscess and paralysis—may or may not have appeared.

For proper examination the patient should be stripped naked. Girls who have reached the aged of puberty and women should receive certain consideration, and it is customary to examine such with the back alone bared. It may be convenient to have the undershirt put on in front as an apron with the sleeves pinned or tied about the neck; the skirts can then be dropped to the level of the greater trochanters and held with a large safety-pin, or by a piece of bandage tied around the hips. The back is then inspected for any deviation, excurvation, incurvation, or prominent vertebræ. If found, the disease may be suspected of being present at the middle of the curvature; but it must be remembered that spondylitis easily demonstrated is usually present some months before deformity of the spinal column is apparent. All of the normal motions should now be tested, both actively and passively; the head should be rotated to right and to left, and the shoulders twisted in the same directions while the pelvis is firmly held; and the spine should be bent forward and backward, and to the right and to the left. Any portion which shows rigidity to all the normal motions is, or has been, the seat of an inflammatory process; but if there be rigidity to bending in one direction only, or if bending in any one direction be normally free, the diagnosis of spondylitis is rendered extremely doubtful. It is upon this rigidity, which for a long time is due solely to involuntary muscular spasm, that the diagnosis must depend; it is ever present, both waking and sleeping, and nothing abolishes it except profound anesthesia, or the termination of the inflammatory process. It is the first symptom to

appear and the last to disappear; and when, and only
when it is no longer present, can a cure be safely
predicated.

Tenderness to direct pressure over the suspected area,

Fig. 4.—Pained facial expression, often seen in patients suffering from
spondylitis.

unless local abscess be present, will not be found. This
local tender point, which is taught by the professor of
and text-book on general surgery as the most important
diagnostic symptom, always counts against rather than

in favor of the diagnosis of spondylitis. It should be remembered that the disease is located in the vertebral bodies, and usually in their anterior parts, and in any case is far beyond the reach of direct pressure; and consequently tenderness to direct pressure, unless supported by strong confirmatory evidence, is to be looked upon as indicative of some other condition than the disease in question.

Fig. 5.—Rigid lumbar spine in commencing spondylitis before the appearance of angular deformity.

Downward pressure and concussion upon the head, and sudden twisting of the spine by wrenching at the shoulders when the patient is off his guard, are tests as unnecessary as harmful. They will not be found to be of any value in the very early period and can scarcely fail to inflict injury as well as pain when the disease is at all well advanced.

Sooner or later deformity of the spine appears, and a lateral curvature, with or without twisting of the vertebræ—rotation—often appears before kyphosis, the so-called "angular curvature," makes its appearance. If but one, two, or three vertebræ be affected, and if the

Fig. 6.—Lateral deformity preceding angular deformity in commencing spondylitis.

destructive process has been considerable, the deformity fairly approximates an angle; but if several vertebræ are diseased each to only a slight degree, the deformity will be a curve.

Motor paraplegia, affecting both lower extremities and at times the bladder and rectum, and at times also the upper extremities, may come on before the bony deformity, or with the bony deformity, or comparatively late in the disease. It is generally due to thickening

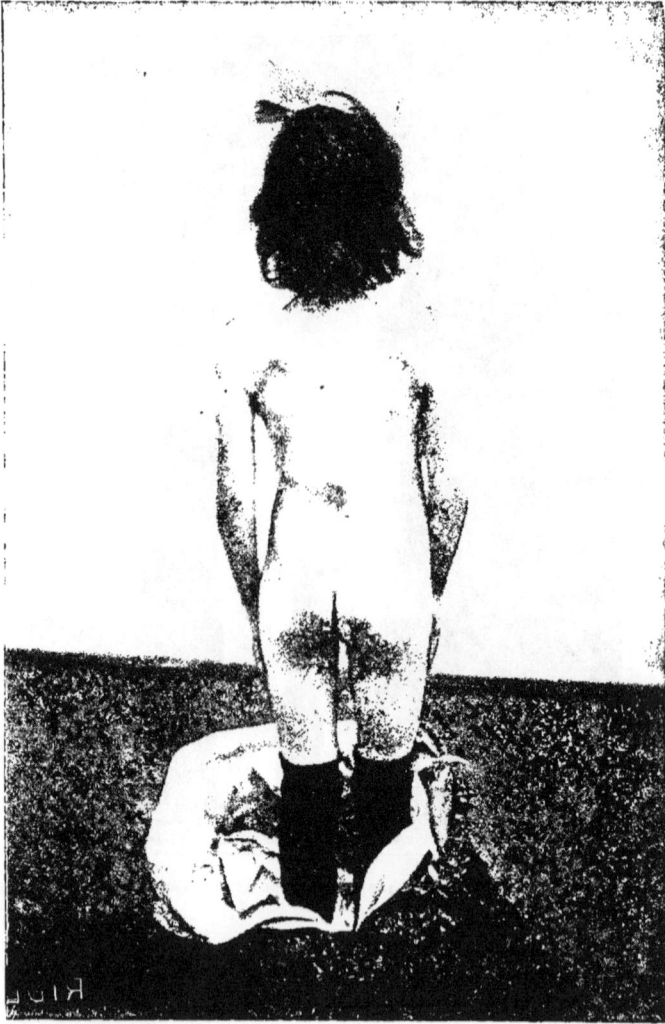

FIG. 7.—Commencing angular deformity of spondylitis.

of the membranes of the cord, from the contiguity of
the inflammation in the bone, occasionally to an
actual invasion by the tubercular inflammation. Paraplegia occurs by far the most frequently when the
disease is located in the upper dorsal region. It bears

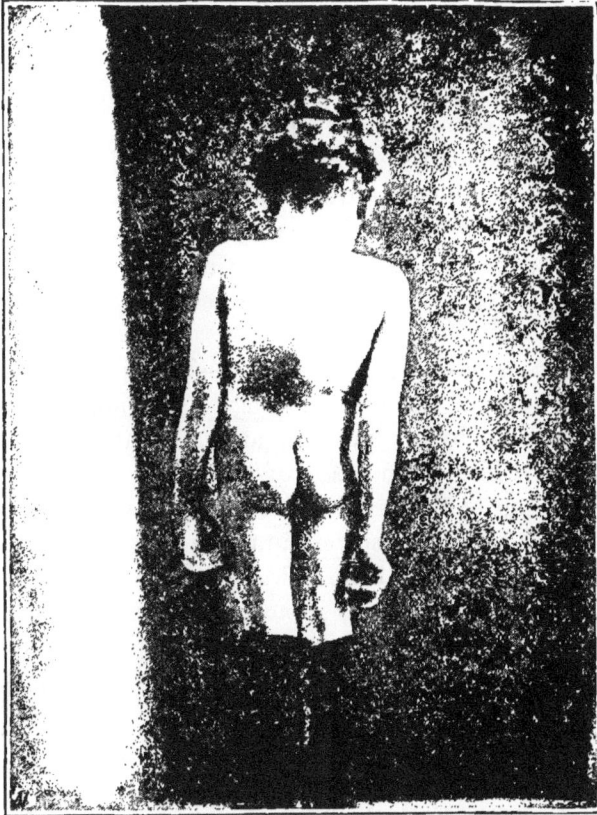

Fig. 8.—Dorso-lumbar spondylitis with shoulders shifted to the right.

no relation to the acuteness of the angle; it may disappear while the bony deformity goes on increasing, and
it has seldom been shown to depend upon bony
pressure. The paraplegia is characterized by an exaggeration of all the tendon-reflexes in the affected

extremities, a tonic spasm of all the muscles, and an inability, more or less complete, to move any portion of the affected parts. The nerves of sensation are very rarely involved.

Although tubercular " pus " is probably formed to some extent in all cases, the tubercular abscess does not appear in more than half the cases. Abscesses are quite frequently seen when the disease is in that part of the spine which lies below the diaphragm; less frequently when disease is in the cervical region; and still less frequently when the disease is in the dorsal spine above the diaphragm. The abscess may make its way in any direction, opening externally, or into any of the open or closed cavities of the body, or it may be absorbed even after it has attained very considerable proportions.

Symptoms of Cervical Spondylitis.—The first symptom is restriction of the normal range of motion; followed after a time by malposition and a greater degree of stiffness. The malposition of the head depends upon the location of the disease. More often than otherwise the upper two or three vertebræ are diseased, and the head is twisted and bent forward and to one side into the position of wry-neck; one or both sterno-mastoid muscles are rigidly prominent, and often the posterior muscles as well. When the disease is lower, the chin is advanced and dropped towards the chest and an angular projection of the spine at the point of disease may be felt; when the disease is still lower, the chin is elevated and relatively somewhat advanced, and the head is thrown backward towards the shoulders, which are raised to meet it. At times the posterior muscles are so much contracted that they simulate an abscess. The face expresses apprehension and the head is moved, if the patient can move it at all, with anxious care. Pain may be complained of running up the back of the neck and head, down the arms, or in the chest. In early cases the deformity, and in fact all of the symptoms, are considerably relieved by even a short period of recumbency.

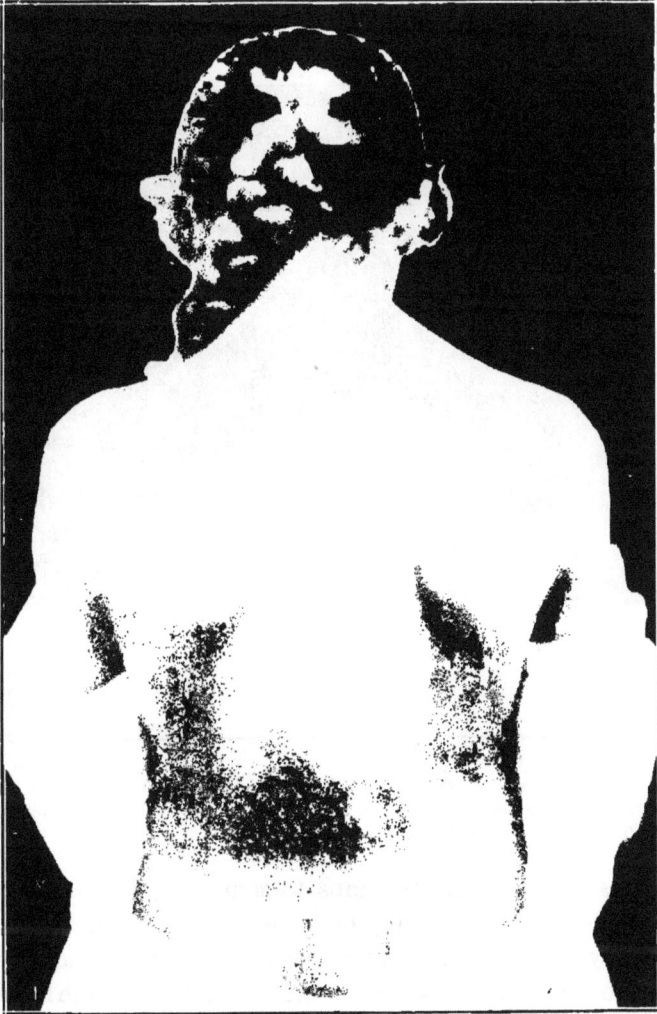

FIG. 9.—An old case of tubercular spondylitis, with posterior, lateral and rotary deformity.

For the proper examination of a case of suspected cervical spondylitis, the patient should be stripped to the waist. The attitude and the range of active motion should be noted; the range of passive motion should be tested, and the neck should be palpated for kyphosis and for a fluctuating swelling. If the range of motion

FIG. 10.—The same case shown in Fig. 9 stooping forward, viewed from the rear, showing by the parallel lines, *a* and *b*, the degree of lateral deviation; showing also the amount of rotary deformity.

be restricted in all directions to some degree, the diagnosis of spondylitis may be considered as certain. In very young children who become frightened at the approach of a stranger the range of motion of the head may be tested by placing the patient across the parent's knees; in the prone position the head will not be let

dangle, no matter how prolonged the examination, and
in the supine position he will not carry it forward as in
the first act of rising.

The complications of cervical spondylitis are abscess
and paraplegia. Abscess does not occur very frequently ;

Fig. 11.—Cervical spondylitis, with twisted head and contracted and prominent sterno-mastoid muscles.

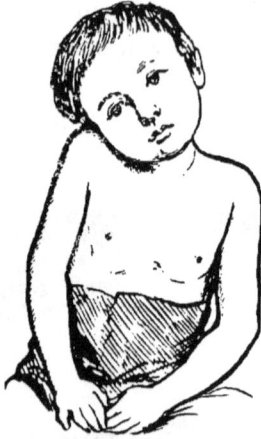

FIG. 12.—Cervical spondylitis; head bent to one side and resting on the raised shoulder.

when it does occur it usually points laterally back of the sterno-mastoid muscle; it may point in the pharynx. The pharynx, however, need not be examined unless some symptom points to abscess in that location. Before the formation of abscess the finger in the throat

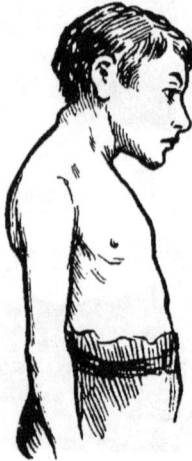

FIG. 13.—Cervical spondylitis; head thrown forward, with chin approaching chest.

will reveal nothing; it is very disgusting to the patient,
and the normal prominence of the vertebral bodies may
mislead the surgeon. Paraplegia is seldom met with.
When present it may affect both upper and lower extremities, or the lower only. A more complete description of the paraplegic symptoms will be given in connection with dorsal spondylitis, of which paraplegia is
more often a complication.

Symptoms of Dorsal Spondylitis.—Before the appearance of kyphosis the diagnosis must depend upon per-

FIG. 14.—Cervical spondylitis; head thrown back; sinus in side of neck
from abscess.

sistent distant pain, often treated for many months as
gastric pains; upon a disinclination to indulge in rough
play; a growing tendency to stand with the elbows
resting on a chair or table; a grunting respiration, and
upon an inability to rise from a stooping posture, or pick
up an object from the floor without resting the hand
upon some piece of furniture or climbing hand over
hand upon his own legs.

Any distant pain which does not readily yield to
proper medication should lead to a careful examination

of the spine. Often there is crying when the child is
lifted, and a cough accompanying a grunting respiration.
If the upper two or three dorsal vertebræ be the ones
involved, the head may be thrown backward and the
neck held rigid to forward or lateral bending, and paraplegia may even come on before any kyphosis can be
made out. When the disease arises in the lower dorsal
region, the patient may limp and complain of pain in
the thigh, as in hip-disease, and this before any deformity is noted. Lateral deviation of the column, with or

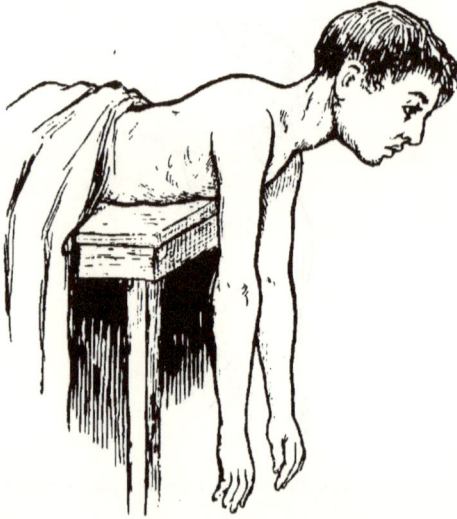

FIG. 15.—Cervical spondylitis. Patient will not let head dangle.

without rotation of the vertebræ, is often present before
the anteroposterior deformity appears.

In patients too young and timid to submit patiently
to an examination by the surgeon it is convenient to
let them lie prone across the parent's separated knees;
if there is disease, the spine will not sag into an anterior
curve in the normal way. If placed sitting upon a table
with the knees straight, the child will not bend forward
arching the spine in the usual way. A spine which is

held rigid in some extent to bending in all directions must be seriously suspected of being diseased.

When kyphosis has appeared, however, the diagnosis will be readily made, for, in addition to the peculiar and striking deformity, all the symptoms heretofore mentioned are likely to be found on careful investigation. Sooner or later projection forward of the chest takes place, compensatory and proportionate to the angular deformity of the back.

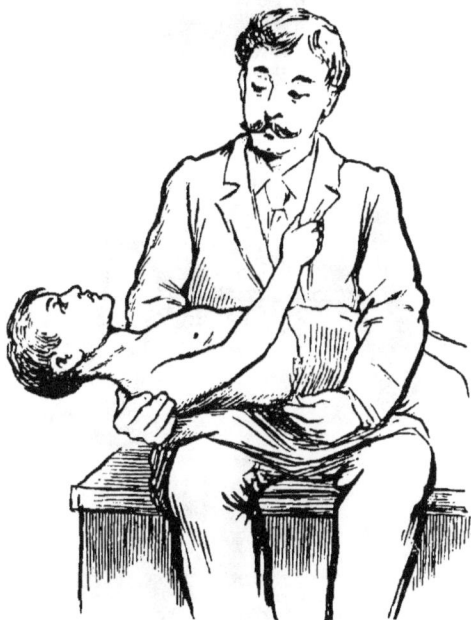

FIG. 16.—Cervical spondylitis. Patient will not let head dangle.

Paraplegia, which is more common when disease is in the upper dorsal spine than elsewhere, may come on early, before any deformity has appeared, or at any time during the course of the affection or during its latest stages, and, having disappeared, may recur again and again. It begins with exaggeration of the tendon-reflexes, stumbling in walking, increasing lack of muscular control, and goes on until all control over the lower extremities is lost and the limbs are held rigidly extended; at times they are drawn up with spasmodic crampings, and may suddenly, without the patient's

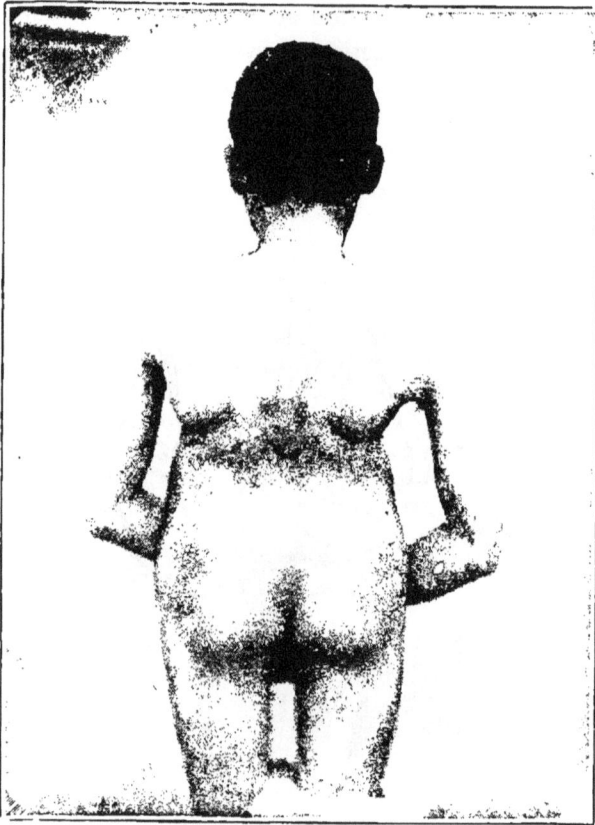

FIG. 17.—Dorsal spondylitis. Great shortening of the trunk; lower angles of scapulæ nearly down to the iliac crests.

volition, be extended with a jerk. Sudden passive dorsal flexion at the ankle-joint induces marked ankle-clonus. On rare occasions the sensory nerves, and at the same time the bladder and rectum, are affected.

Abscess does not frequently appear when the disease is above the diaphragm, although it is probable that collections of tubercular debris form to some extent in all cases. When the abscess does find its way to the surface, it usually makes its appearance from between the ribs at a distance of from one to four inches from the line of the spinous processes; rarely, however, it makes its way downward before appearing at the surface. In

Fig. 18.—Projecting chest accompanying and compensatory to upper dorsal spondylitis.

disease in the lower dorsal vertebræ the abscess usually travels downward in the posterior mediastinum under the ligamentum arcuatum internum of the diaphragm into the sheath of the psoas muscle and thence follows its course into the groin.

Simulating abscess a bursa may form directly over the angular kyphosis in a case subjected to the pressure

!36

of a corset or a corset-brace. We have seen two such cases; in one case the fluctuating tumor was two inches in diameter and in the other three inches. Both tumors were located symmetrically over the most prominent portion of the spine, extending equally to each side of

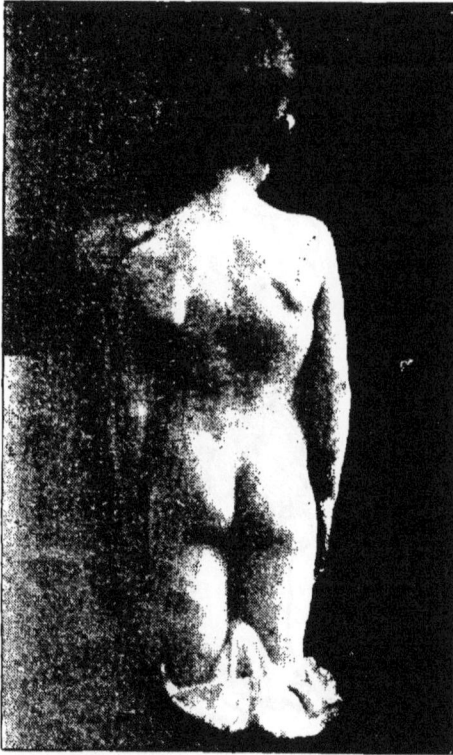

FIG. 19.—Dorsal spondylitis, with abscess below right scapulæ.

the line of the spinous processes. We have not met with a tubercular abscess similarly located.

Symptoms of Lumbar Spondylitis.—The first symptom recognized is usually an awkward gait, a limp, or a slight lordosis. The shoulders are thrown backward, one foot is slightly advanced, and the patient walks

with care and holds his spine rigid. He is even less
inclined than in dorsal disease to forward bending. If
there be pain, it is usually felt down the anterior and
inner surfaces of the thigh. Most of the symptoms
enumerated as characterictic of disease in the dorsal

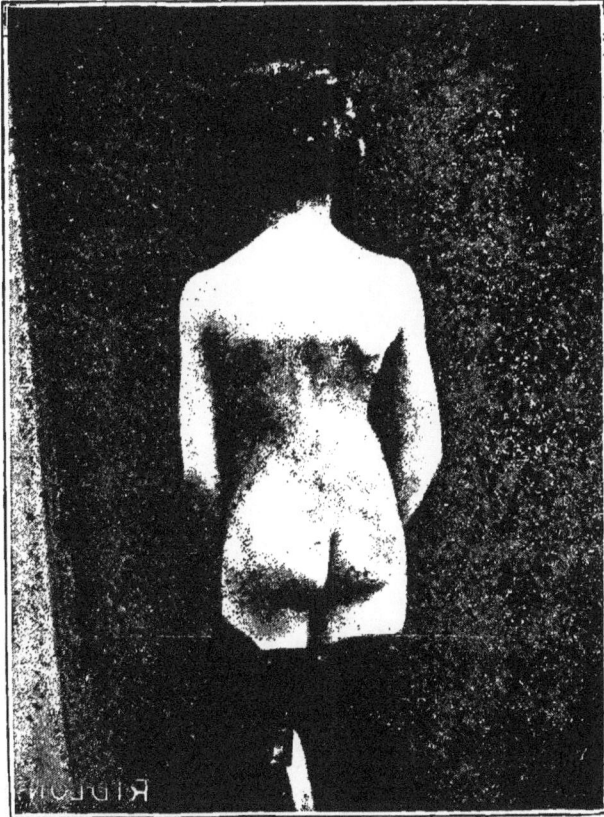

FIG. 20.—Dorsal spondylitis, with sinus from abscess in left loin.

region will be found present. Contraction of one or
both psoas-muscles may come on before the formation
of abscess and before the appearance of kyphosis. It
is this early involuntary spasm of the psoas-muscle,

flexing the thigh and limiting its extension, before the
appearance of deformity that leads to the mistaken diagnosis of hip-disease, even by experienced observers.

To test for contraction of the psoas-muscle the patient
is placed prone upon the table, the pelvis is held firmly

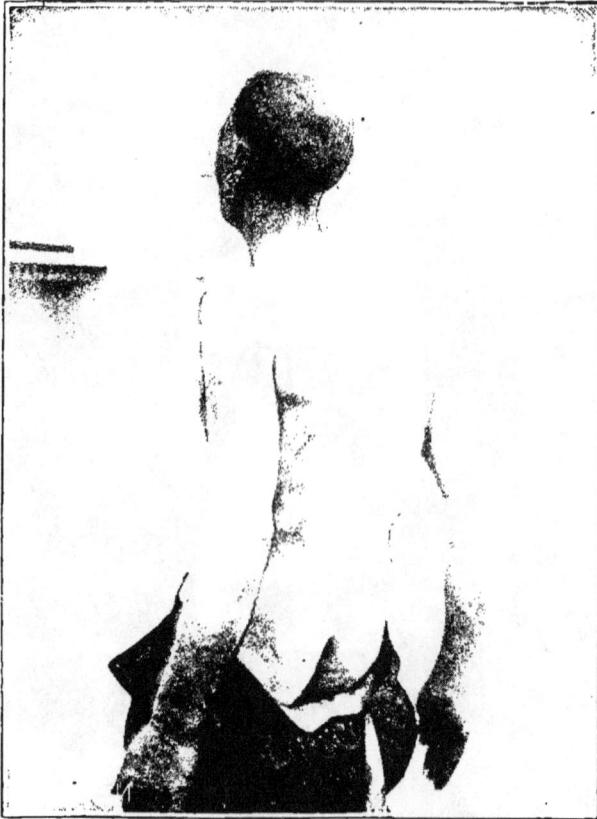

Fig. 21.—The long posterior curvature of dry caries.

down with one hand, while with the other hand, first
one and then the other knee is lifted upwards. The
freedom with which they can be raised and the difference in extent of movement or the extent to which each

of them differs from the normal must be noted. The
rigidity of the spine is tested by placing one hand upon
the back at about the tenth dorsal vertebra, while with
the other hand both of the lower extremities are lifted
at once, carrying the hips with them; in this way not

Fig. 22.—Dorso-lumbar spondylitis, with protuberant abdomen.

only backward bending but also lateral bending can be
tested. It is upon this rigidity that the diagnosis must
depend. In healthy children the spine can be bent
backward so far that the thighs are at nearly a right
angle with the upper dorsal spine.

Paraplegia is not common, owing no doubt to the fact that below the first lumbar vertebra the dura mater only accommodates nerves. The cauda equina does not occupy so great a space as the cord itself along with the nerves leaving it in the higher regions of the spine.

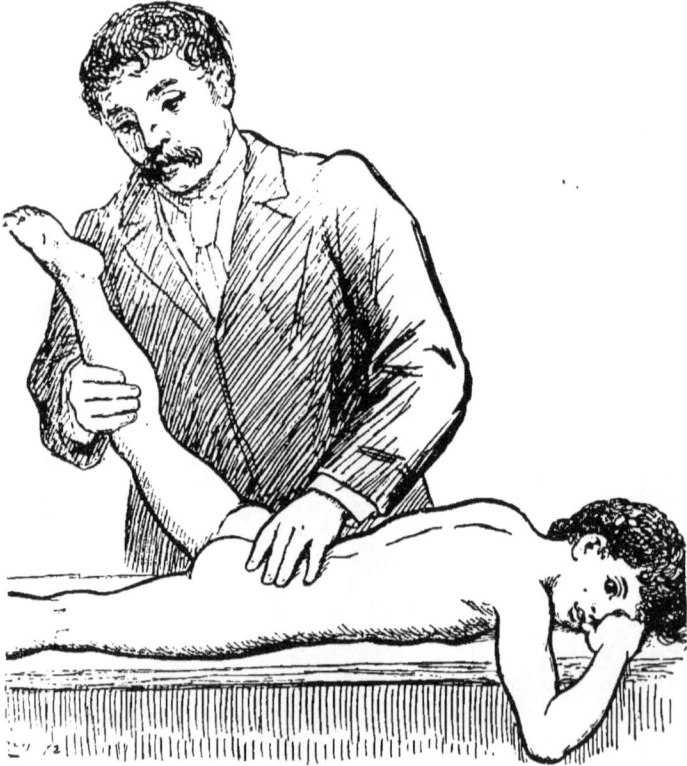

Fig. 23.—Testing for psoas contraction.

When paraplegia occurs, it differs in no way from the same condition when found complicating dorsal disease.

Abscess is frequent, usually following the course of the psoas muscle and pointing on the anterior surface of the thigh below Poupart's ligament, and opposite the insertion of the muscle. When the disease is below

the third lumbar vertebra the abscess may pass down
and point in the buttock; this is due to the entrance of
the pus into the psoas-sheath where it is continuous
with the sacral end of the pelvic fascia, and which
passes down to the pyriformis, and leaves the pelvis
through the great sacro-sciatic foramen ; or with disease
in any of the lumbar vertebræ, the pus may pass laterally, following the nerves, and point in the loin some
inches from the spine.

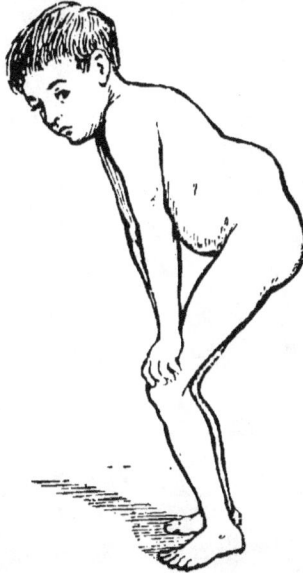

FIG. 24.—Lumbar spondylitis. Patient climbing up his legs after reaching
to the ground.

Differential Diagnosis.—A strain may give rise to the
early symptoms of spondylitis; for a sprain left untreated in a tuberculous subject may become a true
tubercular spondylitis. A strain carries its distinct history of injury, and often presents a local tender point;
the pain on motion is relatively greater and the involuntary rigidity relatively less than is found in spondy-

litis; in a case of sprain there is no angular deformity and rarely any rounded curvature; there is no distant pain, no abscess, and no paraplegia.

Torticollis is closely simulated at first glance by

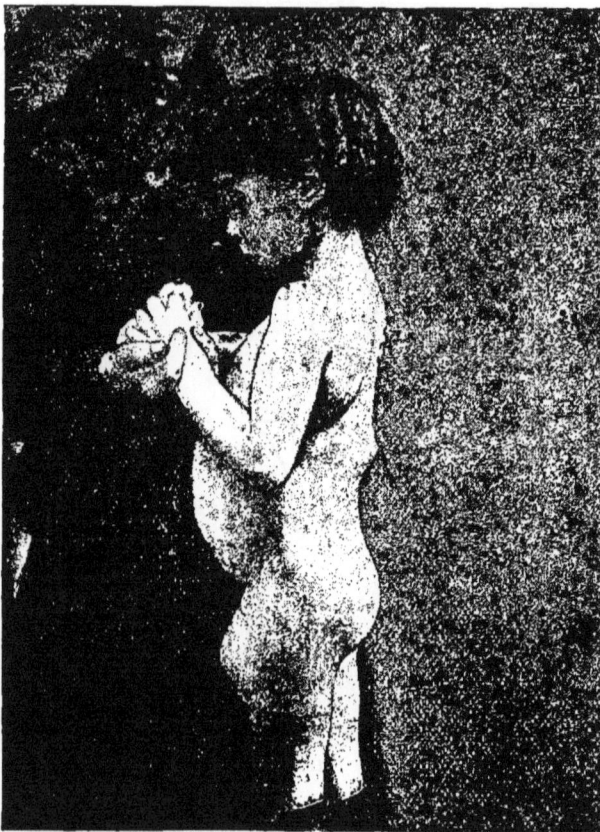

FIG. 25.—Upper lumbar spondylitis with psoas abscess presenting on the thigh in the usual location.

spondylitis in the upper cervical region, and yet, in many cases, the diagnosis can be made at sight. In wry-neck, the chin points away from the prominent sterno-mastoid muscle; in spondylitis, it points towards that muscle, if only one muscle be prominent. In spondylitis, the movements of the head are restricted in all directions; in torticollis, only in one direction— that direction which puts the shortened muscle on the stretch.

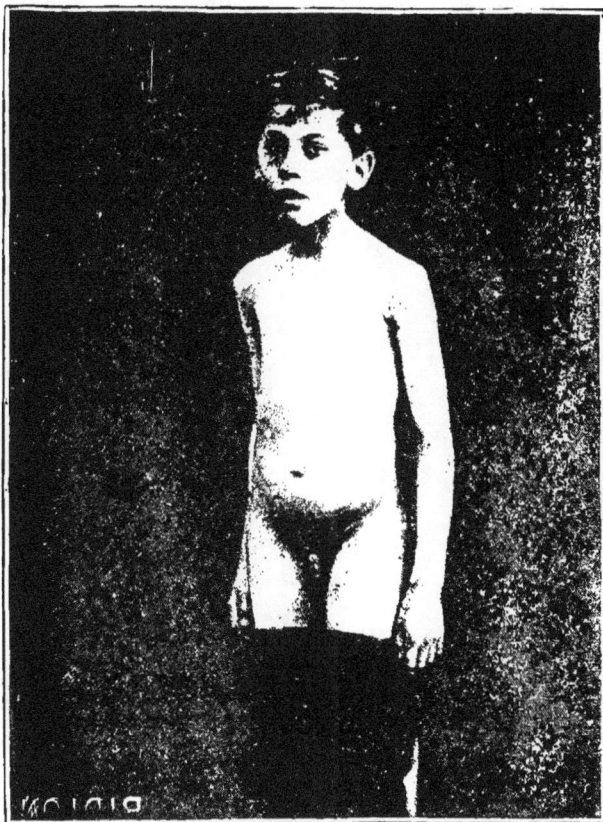

Fig. 26.—Upper lumbar spondylitis with abscess pointing in the groin above Poupart's ligament.

A rachitic spine closely resembles the " rounded curvature" of spondylitis. The deformity is not infrequently a rigid one, but the rigidity has a more elastic feel than that of a tubercular spine; bilateral psoas-

44

contraction is occasionally present. The condition is found only in young children, and is usually associated with other manifestations of rickets. There is no distinct pain, no abscess, and no spastic motor paralysis. The mistaking of a rachitic spine for spondylitis will, however, be of little harm to the patient, since the rachitic spine demands a rigid support.

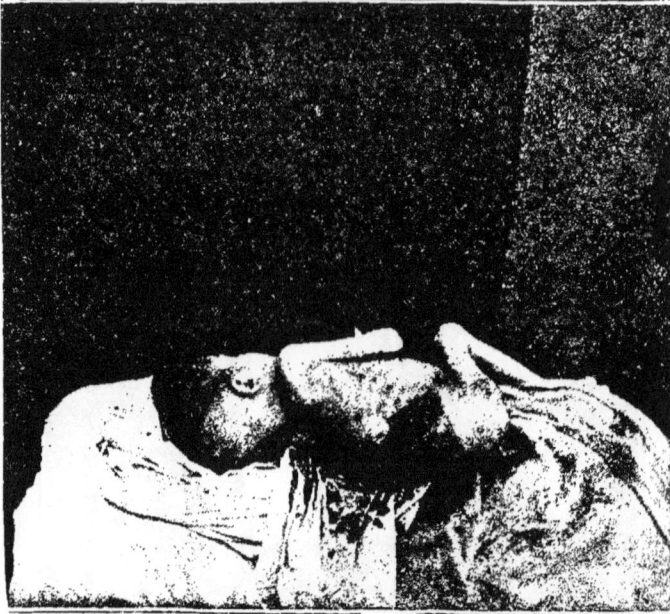

FIG. 27.—Dorso-lumbar spondylitis, infected lumbar abscess with sloughing sinus, enlarged liver and kidneys, and great emaciation.

Scoliosis—rotary lateral curvature—will not be mistaken for spondylitis, as the curvature is not usually rigid until some time has elapsed and the deformity has become considerable. On the other hand, spondylitis may be mistaken for lateral curvature, and the necessary immobilization withheld and possibly exercise advised. A slight lateral curve, with or without rotation, if it be rigid, is probably a commencing spondylitis; exercises should be withheld and a support applied; a few months' observation will clear up the diagnosis. While pain is rarely associated with scoliosis, it is the rule in spondylitis, though it should not be forgotten that it may be absent in both.

The hyperesthetic spine, also called the irritable spine and the hysterical spine, if patiently and care-

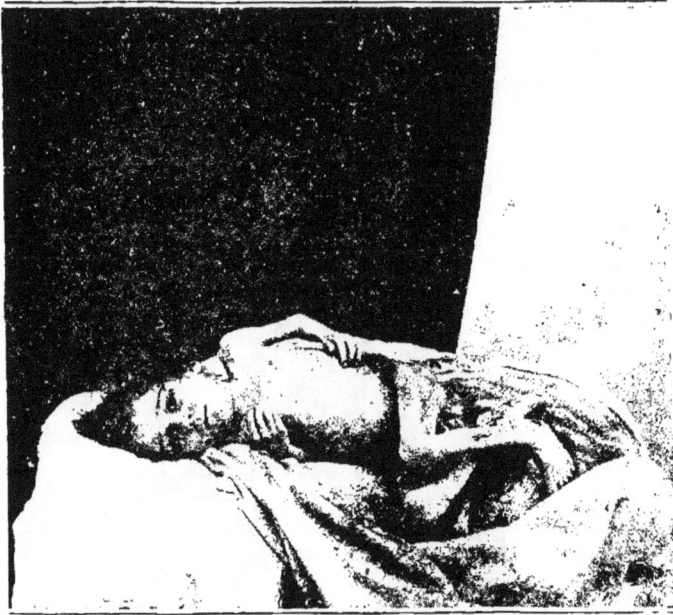

Fig. 28.—Front view of patient shown in Fig. 27.

fully examined, gives no rigidity from involuntary muscular spasm. The spine is held more or less rigid to bending in one direction, or to bending in more than one direction; but the characteristic rigidity of involuntary muscular spasm is wanting, and the patient usually complains of pain on voluntary bending, a symptom very rarely found in spondylitis. There is no distant pain, the pain being confined to some portion of the spine itself and associated with tenderness

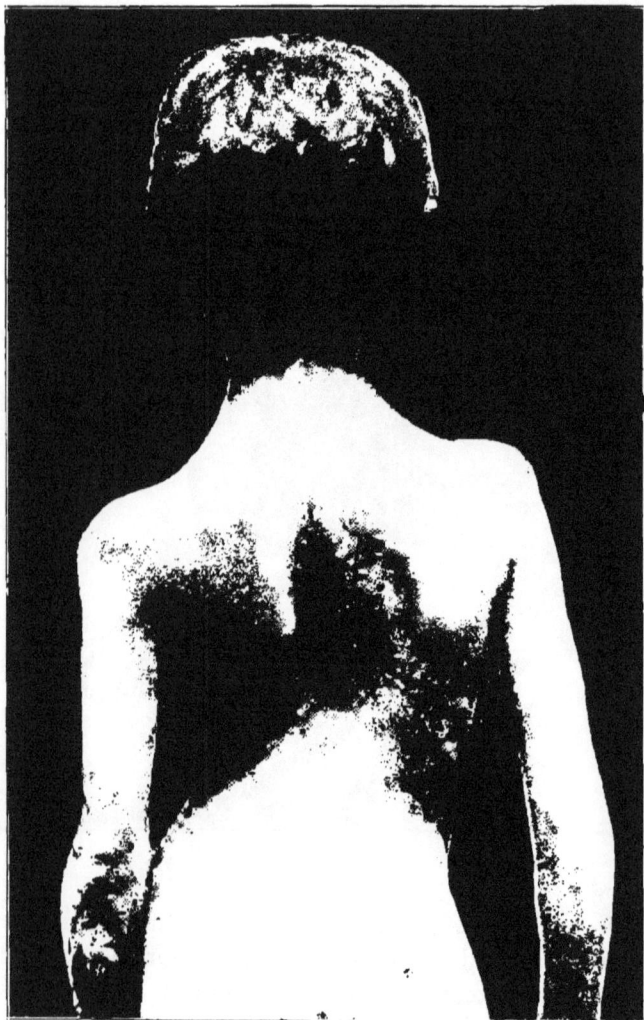

FIG. 29.—A case of tubercular spondylitis simulating scoliosis, showing lateral and rotary deformity. The onset was very rapid and the rigidity great. The diagnosis was not made until treatment by exercises had increased the deformity and rigidity. The deformity was gradually corrected under treatment by immobilization.

on pressure. The condition is most frequently found
in young women, often associated with other hysterical
manifestations, and may remain unchanged for years.
There is no true kyphosis.

Malignant disease of the spine in its very early stage
cannot be diagnosticated from commencing spondylitis.
The history of the case as to hereditary tendency, taken
together with the age of the patient and his general appearance, may make the diagnosis of malignant disease
probable, but nothing can be positively said until the
progress of the disease, or the symptoms due to pressure
of the tumor clear up the doubt. The symptoms in
malignant disease grow steadily worse despite all treatment, while in tubercular spondylitis, treatment by
immobilization and recumbency invariably brings relief
to a very considerable extent.

The typhoid spine can, of course, be found only as a
sequel of typhoid fever. The typhoid spine presents
tenderness on pressure, and on lateral and forward
bending. There is no true kyphosis, no special pain
in the nerve-distribution, and no psoas-contraction;
the onset is sudden, and the recovery rapid.

Hip-disease is not infrequently the diagnosis when
contraction of the psoas muscle comes on in lumbar
spondylitis prior to kyphosis. The patient walks with
a limp, complains of pain in the groin or along the
anterior surface of the thigh, the thigh is flexed on the
pelvis, and attempts to overcome flexion are resisted by
involuntary muscular spasm, and give the patient pain.
It will be found, however, that the thigh can be flexed
to the normal degree, and that when flexed sufficiently
to fully relax the psoas-muscle, rotation at the hip-joint
is free, painless, and normal. In a word, in lumbar
spondylitis extension, and possibly inward rotation
during full extension, are the only motions at the hip-
joint that are restricted by muscular spasm, whereas in

48

hip-disease motion is restricted in all directions. It is
upon this difference in the restriction to motion at the
hip-joint that hip-disease is excluded from the diagnosis.

Sacro-iliac disease is not of frequent occurrence, and

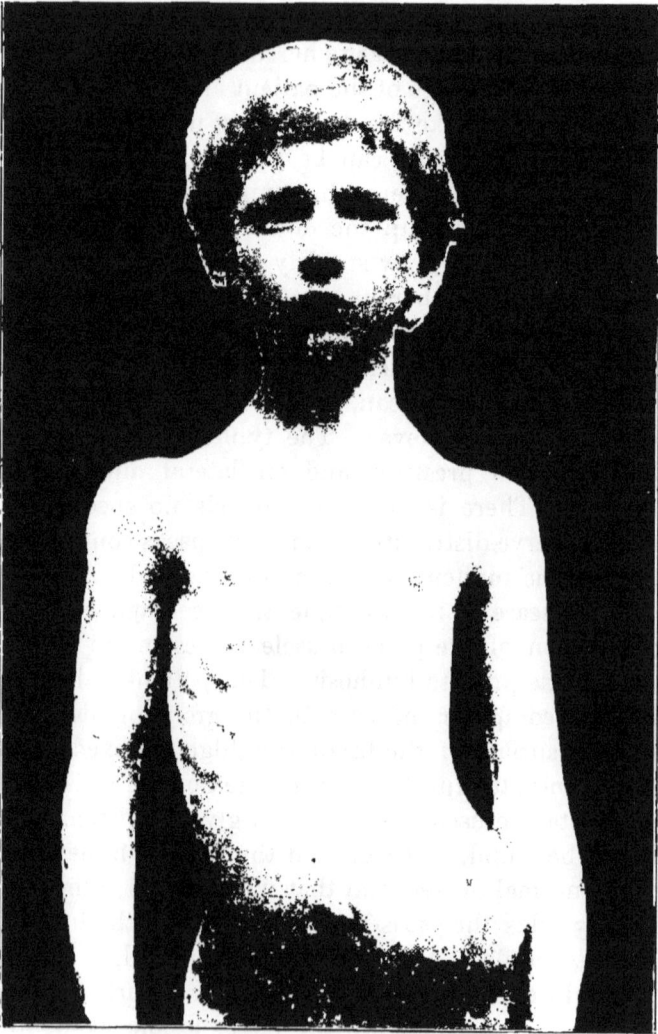

FIG. 30.—Front view of patient shown in Fig. 29.

its early symptoms are obscure. It rarely occurs in young children, except in association with disease at the sacro-lumbar junction. When the sacro-iliac joint alone is involved, the lumbar spine will be held rigid on forward bending, but gentle, passive bending backwards and laterally will generally be found to be free. In all suspected cases examination should be made by way of the rectum, since swelling appears earlier within the pelvis than upon the surface external to the joint.

The prognosis of spondylitis must be considered as to deformity and function, as to duration, as to life, and as to the complications.

As to deformity and function : In the cervical and dorso-lumbar regions, under favorable opportunities as to treatment, the deformity may be reduced, and often entirely eradicated, if consolidation has not already taken place, and increase in the deformity may be prevented even if consolidation has commenced. In the upper dorsal region, from the first to the sixth vertebra, the deformity may be expected to increase under any form of treatment which does not include prolonged and uninterrupted recumbency as its essential feature. From the sixth to the tenth dorsal vertebra an increase can generally be prevented, but rarely can the deformity be reduced except by forced straightening under an anesthetic. When the disease affects the lower lumbar region, the fourth and fifth vertebræ and sacrum, the deformity may be expected to come on, and to increase up to a certain point, unless the patient be treated continuously in the recumbent posture until consolidation is well advanced. In a word, if the spine can be made straight and kept so sufficiently long for the ossific matter to deposit in the space made vacant by the disease, an ankylosis, free from deformity, or nearly so, will result. In a few cases, more or less, restoration of the normal motion is gained. The lack of early diag-

50

nosis, and of early, energetic and prolonged treatment,
may be considered as the cause of the deformity which
ultimately results in so many cases.

As to duration : It must not be expected that any

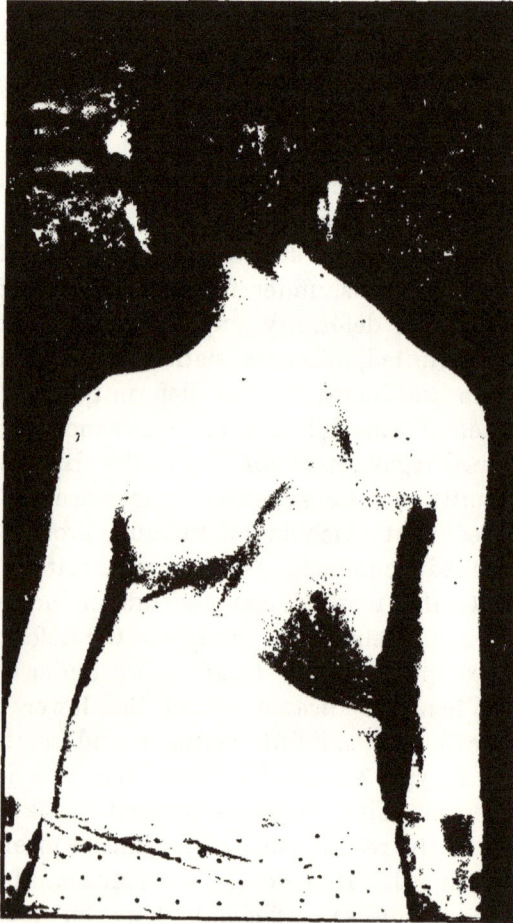

Fig. 31.—Upper dorsal spondylitis with marked lateral and rotary deformity. The development of the deformity was comparatively rapid ; it was very
rigid ; and horizontal traction by weight-and-pully, in bed for a year, very
greatly reduced the deformity, diminished the rigidity and confirmed the
diagnosis.

case, even the one presenting only the few symptoms requisite for a certain diagnosis and most favorably circumstanced as to nursing and treatment, will recover within a year; two years constitute a short time, and the average case will require treatment for from three to four years, and many a much longer time.

As to life: Although spondylitis is a most prolonged and serious disease, the prognosis as to life is remarkably good. In neglected cases the death-rate may run as high as 30%; in cases receiving proper nursing and conservative treatment the death-rate is not more than

FIG. 32.—Cervical spondylitis simulating wry-neck. The chin points towards the prominent sterno-mastoid muscle. In wry-neck the chin points away from the prominent sterno-mastoid muscle.

8%; to this may be added 10% for such cases as are subjected to operative measures.

As to the complications : Abscesses are reabsorbed in about half the cases, provided the spine is properly protected. Abscesses that open spontaneously and are left to empty without interference heal as kindly as sinuses resulting from incision, and rarely afflict the patient with symptoms of septic poisoning. Abscesses that are incised rarely heal by first intention, and as a rule become septic and continue to discharge as long and often longer than those left without interference. The duration of an abscess arising from tubercular spondylitis covers months and often years.

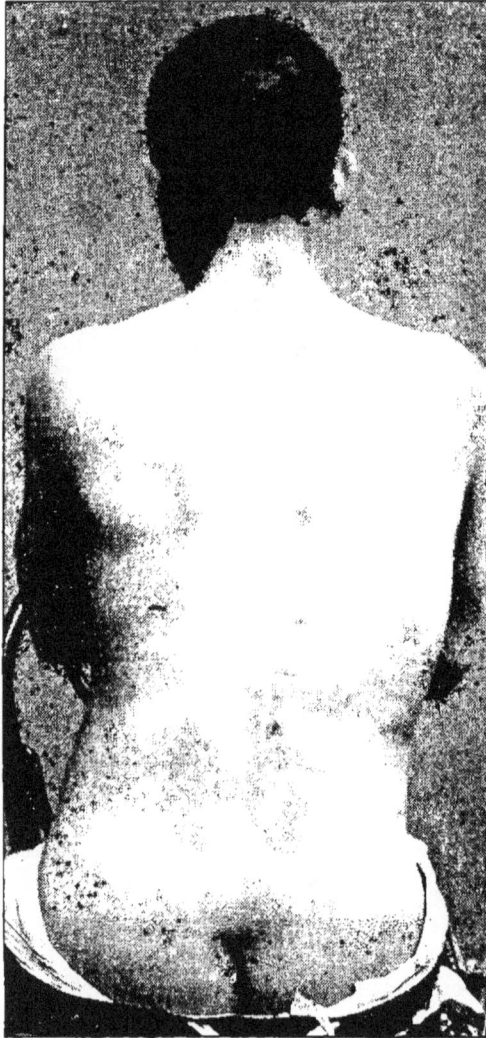

FIG. 33.—A hyperesthetic spine, with lateral deviation, simulating commencing dorsal spondylitis.

Paraplegia, being due in nearly all cases to pressure upon the cord from the inflammatory products in the neighborhood of the bony tuberculosis, may be expected to pass off as the active inflammation subsides; all such cases recover from the paralysis, if the patient be kept recumbent sufficiently long. A few cases die from an invasion of the cord itself by the tubercular inflammation and a few from a displaced bony sequestrum so placed as to press upon the cord. In all cases in which the sensory nerves and the bladder and rectum are involved a fatal issue may be anticipated. The average duration of the paraplegia in cases subjected to conservative treatment is somewhat more than thirty weeks; but the authors have observed a complete restoration of function to the paraplegic limbs in a case where the motor paralysis had been complete for nearly four years, and another case in which there has recently been a partial recovery where the sensory paralysis had existed for two years and the motor paralysis for ten years. Recent work by various surgeons in forcible straightening of kyphotic spines under anesthesia seems to warrant the expectation that this operation will give immediate relief from the paralysis in the vast majority of cases. Relapses of the paraplegia may occur, but fortunately are not frequent.

The treatment of spondylitis, like the treatment of chronic disease of the arms and legs, is chiefly mechanical, but occasionally it is operative. The mechanical treatment may be divided into three stages: 1. Correction of the deformity; 2. Physiologic rest of the diseased area; 3. Restoration of function.

Correction of the Deformity.—From the earliest times efforts have been made to correct spinal deformities. Hippocrates (460–357 B. C.) treated spinal deformities occasionally by succussion, but he appears generally to have preferred treatment by forcible traction and countertraction, with direct pressure upon the gibbosity.

Henry Heather Bigg,[1] quotes the following from Dr. Adams' translation of "The Genuine Works of Hippocrates ":—

"Those cases in which the gibbosity is near the neck are less likely to be benefited by these succussions with the head downwards, for the weight of the head and tops of the shoulders when allowed to hang down is but small; and such

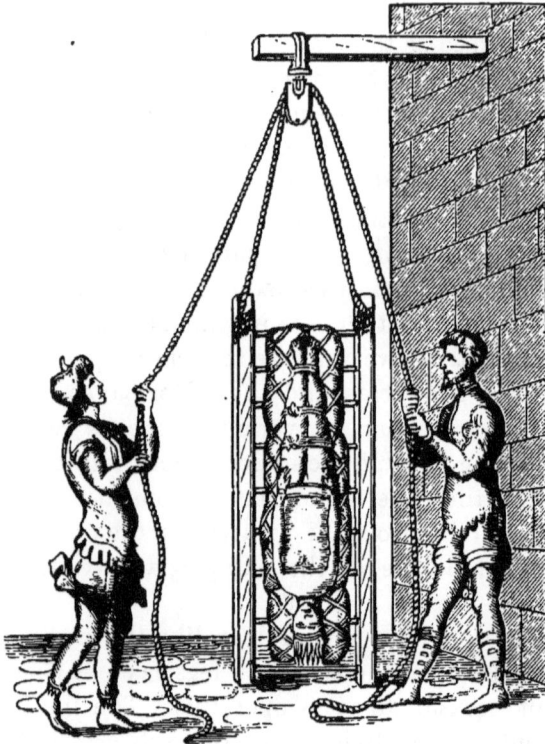

FIG. 34.—The treatment of spinal deformities by succussion. From the Venetian edition of Galen. Quoted in Bigg's "Orthopraxy."

cases are more likely to be made straight by succussion with the feet hanging down, since inclination downwards is greater in this way. When the hump is lower down it is more likely that succussion with the head down should do good. If one should think of trying succussion it may be applied in the

---

[1] "Orthopraxy." J. & A. Churchill, London, 1877.

following manner : The ladder is padded with leather or linen cushions, laid across, and well secured to one another, to a somewhat greater extent, both in length and breadth, than the space which the man's body will occupy ; he is then to be laid on the ladder upon his back, and the feet at the ankles are to be fastened, at no great distance from one another, to the ladder with some firm soft, cord; and he is further to be secured in like manner both above and below the knee and also at the nates; and at the groins and chest loose shawls are put around in such fashion as not to interfere with the effects of succussion; and his arms are to be fastened along his sides to his body, and not to the ladder. When you have arranged these matters thus you must hoist up the ladder, either to a high tower or to the gable end of a house; but the place where you make the succussion should be firm and those who perform the extension should be well instructed so that they may let go their hold equally to the same extent, and suddenly, and that the ladder may neither tumble to the ground on either side nor they themselves fall forward.

FIG. 35.—Forced correction of spinal curvature by traction and countertraction, and direct pressure by lever. From the Florentine edition of Galen. Quoted in Bigg's " Orthopraxy."

But if the ladder be let go from a tower, or the mast of a ship fastened to the ground with its cordage, it will be still better, so that the ropes run upon a pulley or axle-tree."

For the treatment of the gibbosities of spinal caries by extension Hippocrates recommended that

" something like an oaken bench of a quadrangular shape is to be laid along at a distance from the wall in which a groove has been previously scooped, which will admit of persons to pass around if necessary, and the bench is covered with robes, or any thing else which is soft, but does not yield much."

The patient after being stoved and bathed with hot water is to be stretched upon the board on his face, the arms being laid along and bound to his body. Next

"the middle of a thong which is soft, sufficiently broad and long, and composed of two cross straps of leather, is to be twice carried along the middle of the patient's breast, as near the armpits as possible; then what is over of the thongs at the armpits is to be carried round the shoulders and afterwards the ends of the thong are to be fastened to a piece of wood, resembling a pestle; they are to be adapted to the length of the bench below the patient, and so that the pestle-like piece of wood resting against this bench may make extension. Another such band is applied above the knees and ankles, and the ends of the thongs fastened to a similar piece of wood; and another thong, broad, soft, and strong, in the form of a swathe, having breadth and length sufficient, is to be bound tightly around the loins as near the hips as possible; and what remains of the swathe-like thong with the ends of the thongs must be fastened to the piece of wood placed at the patient's feet, and extension in this fashion is to be made upwards and downwards, equally, and at the same time in a straight line."

It is further recommended to press the palm of the hand upon the hump while extension is being made; or a person may sit upon the hump while extension is being made, rising from time to time and letting himself fall back upon it, or the foot may be placed upon the hump and the entire weight of the body brought gradually to bear upon it. Or better still, a lever may be used, one extremity of which is fixed in a hole in the wall, or in a piece of wood fastened to the ground. This lever is brought across the hump, a cushion being interposed and firmly pressed down while extension is made.

Galen (130–200 A. D.) appears to have followed very closely the methods of Hippocrates. His influence was paramount for more than 1300 years. In the Venetian edition of Galen's works will be found an illustration showing the method of performing succussion. In the Florentine edition is an illustration showing pressure upon the gibbosity by means of a lever during extension and counterextension by a windlass device.

Ambroise Paré (1517–1590) followed the teaching of Hippocrates in all essential particulars. He differs only in setting aside the pestle-like lever of Hippocrates and the windlass device of Galen for direct manual traction and countertraction, and in giving certain instructions for exerting pressure upon the projecting portion of the spine. He adds, moreover, directions for the application of splints to the back when the distortion has been reduced. The illustration that he gives shows the patient laid upon his face on a table, bound with towels under the armpits and about the hips, and by means of these extension is made, but not violently. During the extension pressure is made upon the projecting vertebræ by the hands; but if pressure exerted in this manner fails to restore the protruded parts, then it

"will be convenient to wrap two pieces of wood, four fingers long and one thick, more or less, in linen cloths, and so to apply one to each side of the dislocated vertebræ, and so with your hands to press them against the bunching forth vertebræ until you force them back into their seats, just after the manner you see it before delineated." ("Orthopraxy." Henry Heather Bigg. 1877.)

These ancient methods were long since abandoned and have only been saved from total oblivion by occasional mention as examples of curious and barbarous procedures. Reference is here made to these methods for comparison with the work recently done by Calot and others, to be described later on in these pages. In 1874 Sayre, of New York, began the treatment of spondylitis with the use of the plaster-of-Paris jacket,

58

applied with the patient partially suspended, claiming
by that means to correct the deformity in some considerable measure. Orthopedic surgeons in general,
however, denied that the true curve or angle at the

Fig. 36.—Correction of spinal curvatures by traction and countertraction and manual pressure upon the kyphosis. From Ambroise Paré, in the sixteenth century. Quoted in Bigg's "Orthopraxy."

point of disease was in any way affected by this treatment, and that the only straightening of the spine
effected by it occurred in the compensatory curves
above and below the area of disease. The recent results
from forcible straightening of spinal curvatures seem

to demonstrate that Sayre's early claims were well founded.

Charles Fayette Taylor, of New York, followed with claims of gradual straightening by the antero-posterior leverage spinal brace; and his son, Henry Ling Taylor, has shown tracings of many cases treated with the Taylor brace and a more or less prolonged period of recumbency, in which some degree of straightening has taken place.

Following the treatment recommended by the late Dr. Buckminster Brown, of Boston, we believe that one of us (J. R.) was the first to report a case of cervical spondylitis in which a well-marked angular deformity was completely corrected by horizontal traction, the patient being in bed.

The attempts at gradual straightening by the leverage-brace, and by traction, with the patient recumbent, have been successful so many times and in the hands of so many different men that we may speak very confidently as to the results. They are briefly as follows : So long as the disease is active, deformity from disease in the cervical spine can be wholly effaced; deformity in the upper half of the dorsal region can rarely be reduced; in the lower half of the dorsal region it can usually be somewhat reduced; in the lumbar region it can usually be greatly reduced and sometimes entirely effaced.

Recently, efforts have been made, and successfully, to straighten these spinal deformities by the exercise of considerable force, the patient being completely anesthetized. For many years, since the introduction of ether and chloroform, surgeons have been accustomed to straighten deformities due to disease at the joints of the legs and arms by the use of considerable force, but to Dr. Calot, of Berck-sur-Mer, is due the credit of suggesting and employing the same procedure in deformities due to disease in the spinal bones. Calot says that he was not satisfied with the results of the usual methods of treating Pott's disease; that he found the deformities under the usual methods of treatment growing progressively worse. This same experience has been the common lot of all surgeons who depend upon a surgical instrument-maker to measure for and apply spinal braces, and it has too often been the experience of those depending upon the plaster-jacket as a support in ambulatory cases.

Calot has been followed, with certain modifications, by very many of the Continental surgeons. It is not necessary here to specify them or their modifications. Briefly they are as follows: In recent cases the straightening is effected by longitudinal traction by the hands of assistants or by mechanical devices, while the operator makes downward pressure upon the kyphosis, the patient lying prone. In older cases operative measures are added. The soft parts are cut through and the bones divided by a chisel in one or more places; carious foci, if within reach, are scraped out; then the spine is straightened, and the spinous processes are wired to each other to maintain the correct position. This idea of immobilization by wiring the spinous processes appears to have originated with Dr. B. Hadra, of Galveston, Texas, who, in 1890, wired the spinous processes in a case of fracture of the spine. The immediate result was so good that Hadra recommended it in the treatment of Pott's disease in a paper read before the American Orthopedic Association in 1891. Hadra's wiring for fracture ultimately proved a failure, and, so far as we know, it has not been attempted for Pott's disease in this country. European surgeons appear to be becoming more conservative in their work in forcibly straightening these spines, since deaths have occurred from tuberculous meningitis and general tuberculosis arising apparently from dissemination of the disease through tearing through the walls of the tuberculous focus, as has occurred after forcible straightening at the hip and knee.

At the present time the deformity, if it be at all considerable, is straightened at several operations instead of at one, and cutting operations are avoided. Plaster-of-Paris is generally used as the means of retention. It may be put on while the patient rests upon two blocks, one under the hips and the other under the upper part of the chest, traction and countertraction being maintained; or the patient may be suspended by the feet during the application of the jacket. Foot-suspension requires fewer assistants and is safer when chloroform is used as the anesthetic; the objection to it is that the abdominal contents are displaced upward to an unnatural and uncomfortable degree. The horizontal position is objectionable chiefly because it favors an uncomfortable degree of lordosis when the disease is dorsal, and because it is a difficult position when the head needs to be embraced by the plaster-dressing. Unless the head be included in the dressing, recurrence of the deformity may be expected in all cases in which the disease is at or above the ninth dorsal vertebra. After forcible correction of the deformity the patient must be kept in bed for many months, as no mechanical device can be relied upon to retain the spine in the corrected position if patients are allowed to walk around.

We have performed this operation of forcible straightening a sufficient number of times to warrant us in speaking from personal experience. The patient is fully anesthetized, perferably by chloroform. The patient is then turned prone upon the table. One assistant, in the case of a child, makes traction upon the legs; another makes traction upon the arms, and the anesthetizer makes gentle traction upon the head. The operator, with his hands upon the kyphosis, directs the gradual increase of the traction as the deformity yields, usually with considerable tearing and crackling, which can be distinctly felt and sometimes be heard. In recent cases little or no pressure need be made by the operator to effect the straightening of the spine; in cases of longer duration the surgeon may find it necessary to exert very considerable pressure. Each operator must judge of the amount of force that can be safely employed in each case. When the deformity is in the dorsal region and associated with a protruding sternum it will be convenient to place a block under the hips and another under the sterno-clavicular articulation. When the straightening has been effected the patient is clothed in a closely fitting stockinet shirt; quarter-inch felt pads must be placed over the iliac crests, and from half-inch to inch pads of felt must be placed upon the transverse processes on each side of the kyphosis, as close as possible to the spinous processes, to guard against press-ure-sores, and then the plaster-jacket is made. Such a jacket must be made longer than the ordinary plaster-jacket used in ambulatory cases. In cases of lumbar disease it must extend downward on the thighs below the greater trochanters so far that the patient cannot sit upright, and in the presence of disease at or above the ninth dorsal vertebra it must extend up to and include the head. It has been claimed that the jacket thus applied maintains the longitudinal extension, but this is doubtful; it does, however, almost wholly pre-vent antero-posterior bending, and thereby acting as a lever, it prevents any considerable return of the deformity. The application of a plaster-jacket, including the head, while the patient is profoundly anesthetized and horizontal traction and countertraction are main-tained, requires many assistants and is rather difficult; if sufficient assistants are not available it will be best to suspend the patient by the feet, or by the knees, and apply the jacket with the head pendant. When patients are to be suspended by the feet or the knees plaster-casts must be put on the feet, or the knees, the day before the operation, so that the supporting straps or bandages may not constrict the limbs. When the head is to be included the hair is to be cut short, or the head shaved, and then wrapped thickly in bandages of cotton-wadding or wrapped in oakum held by an ordinary roller-bandage. Pressure-sores readily form on all parts of the head and they may form over the prominent sternum, or indeed over any bony point.

After observing the recent work of Goldthwait, of

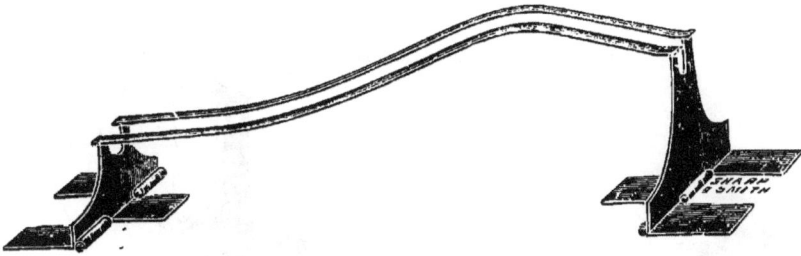

Fig. 37.—Ridlon's Bridge for supporting a patient during application of plaster-jacket.

Boston (May, 1898), who utilizes the weight of the patient above the diseased area to partially correct the deformity, without anesthesia, and his stretcher-frame for holding the patient during the application of the plaster jacket, one of us (J. R.) has made use of the following procedure: After the spine has been straightened forcibly, as already described, and the patient is clothed in the stockinet shirt, he is laid supine upon two light steel bars supported by two sheet-steel rests that stand upon a table. The steel bars are bent to fit the straightened spine from the apex of the kyphosis downward, and they are separated just far enough to make pressure upon the transverse

processes. The sheet-steel stands that carry the parallel
bars are narrowed at the top to $\frac{1}{8}$ inch on each side of
the bars. Laid supine upon these bars, with the part
of the body above the kyphosis extending beyond
their ends, the weight of the upper part of the
body will straighten the deformity more than it can be
straightened in any other way during the application
of the jacket. A half-inch pad of felt is placed between
the bars and the kyphosis, and quarter-inch pads over

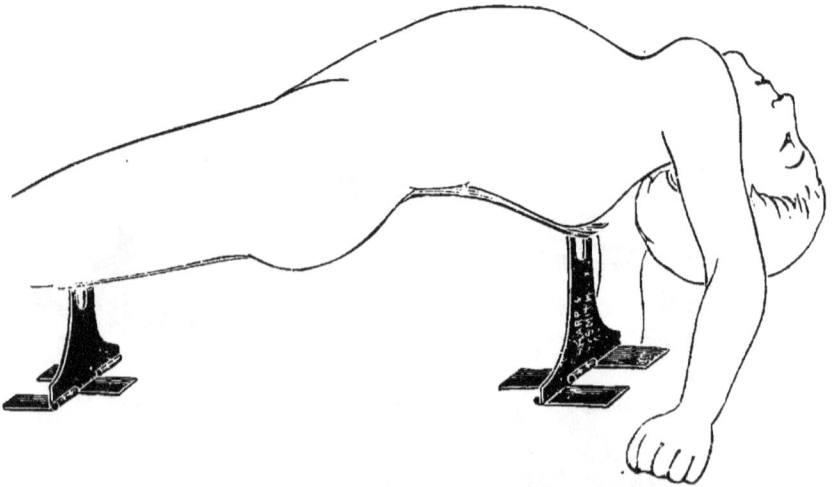

Fig. 38.—Patient resting on Ridlon's Bridge in position for the application
of a plaster-jacket.

the iliac crests and over the upper part of the sternum.
The plaster-jacket is then made, including the parallel
bars. When it has hardened the patient is turned
over prone, and the parallel bars are pulled out. This
leaves a weak place at the back just above the angle of
the deformity; and this can be strengthened by a few
half turns of a plaster-bandage. In cases in which the
disease is in the dorsal region above the ninth vertebra, the jacket can be built up in front under the chin
of the extended head and the whole head need not be

included in the plaster-dressing. This bridge-device
for supporting the patient, which only weighs 2 or 3
pounds, can be used in place of Goldthwait's stretcher-
frame in applying plaster-jackets when the patient has
not been anesthetized, and in all cases it is far more
convenient for the surgeon and more comfortable for
the patient than any form of suspension.

The second stage of treatment may be summed up
in the term physiologic rest. This means the nearest
possible approach to immobilization of the diseased
area, its protection from jars and concussion, and its
relief from weight-bearing until consolidation has become well established. Immobilization is sought for,
and more or less perfectly attained by rest in bed, with
or without traction, and by the application of a brace
or corset; some of these devices by their leverage-
action protect the diseased area from a certain degree
of weight-bearing and jar during locomotion. Whether
the patient should be treated with brace or jacket, or
in bed with or without traction, will depend not only
upon the individual case and its personal peculiarities,
but also upon the family and the general surroundings,
and the skill of the surgeon himself in the use of this
or that special appliance. One surgeon may be able to
fit a brace well and manage it skilfully, but be unable to make a satisfactory plaster-jacket; another may
not be able to work at all well with tools and yet make
an elegant and well-fitting jacket. The methods of
treatment are of far less importance than a correct ap-
preciation of the principles involved, and the patience
requisite to carry them out to the very end. In the
hands of one of us the best results have been obtained
by the antero-posterior leverage-brace; in the hands of
the other a cuirass has been most serviceable. The
latest form of the Taylor brace is perhaps the most correct theoretically, but in our hands it is not readily ob-

tained and fitted. Most surgeons probably now use some form of jacket of plastic material, of which the Sayre pattern is the example best known and most readily made. Each of these will be hereafter described.

Treatment should be commenced at the earliest possible moment, and must be persisted in until cure is effected. A case is cured only when the spine will bear the superincumbent weight without pain or evident weakness in any posture, and continue so without any increase in the deformity.

Treatment by Recumbency.—This mode of treatment, by recumbency, in its effective application, is so exacting to the patient as to be well-nigh impossible. It calls for the most careful nursing and hence is wholly unsuited for the poorer population. It requires that the bed should be flat, smooth, firm, and without a pillow, and the patient so secured by straps that he cannot sit up, twist or turn. Thus, a strap of webbing or strong bandage is passed across the bed beneath the patient's shoulders, and fastened to the bed-frame on either side; upon this strap are strung two loops through which the patient's arms pass, and these are connected the one with the other by a strap across the chest. The pelvis is secured by a broad belt around it, from the sides of which straps pass to the sides of the bed-frame and are there fastened. The patient must not be allowed to sit up for food, for the use of the bed-pan, or for any other purpose; nor must he be taken from bed for bathing, for changing of sheets or clothing, or for the turning of the mattress, if the best effects of recumbency are to be secured. To fail in strictly following these directions will cause the breaking up of the new bone-formation about the carious vertebræ, a return or an increase of the deformity, or it may prolong the paraplegia, if present, and perhaps render it incurable. It will be readily seen that although the

surgeon is saved labor, it is very difficult to carry out
this treatment for any considerable time; in fact, practically impossible to carry it to a successful result in

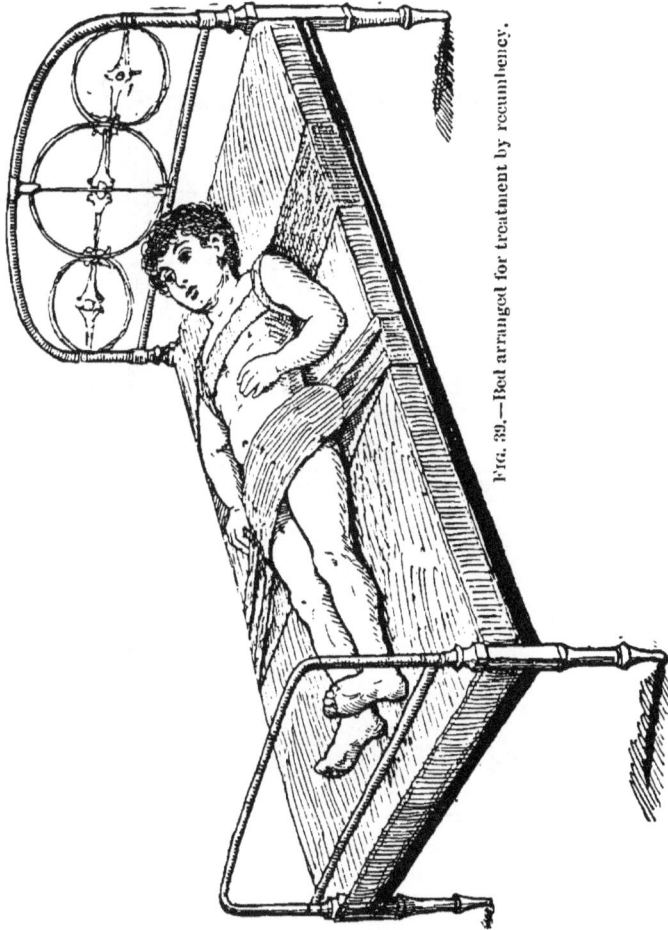

FIG. 39.—Bed arranged for treatment by recumbency.

any but exceptional cases. So-called "treatment by
recumbency" often means that the patient lies in bed
when he chooses, sits up when he pleases, or gets up
and walks when he can. Under such conditions it is

not surprising that the deformity increases, that abscesses are frequent, and the duration of the disease is prolonged.

To increase the efficacy of recumbency various other means have been employed. Traction is used, both

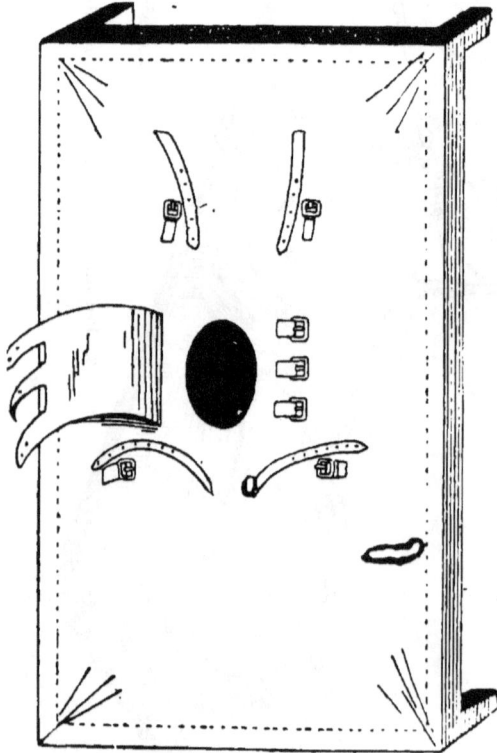

Fig. 40.—Canvas cot for treatment by recumbency.

for its effect in reducing the deformity and preventing the patient from sitting up in bed. A sling is attached to the patient's head and from it a cord is carried over a pulley at the head of the bed to a small weight of from $\frac{1}{2}$ to 1 or 2 pounds. Slings may be made of leather or of cloth, and are constructed upon one or two general

plans; either that which is used in the ordinary suspension-apparatus, or one made by buckling a band
across the forehead and below the occiput and attach-
ing another above each ear to pass over the head, to
which is attached the pulley-cord. This leaves the
chin free, prolonged use not causing recession of the

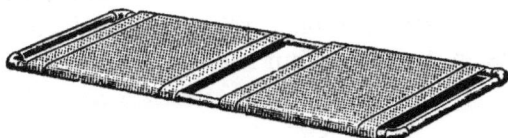

Fig. 41.—Bradford's frame for treatment by recumbency.

chin, and in exquisitively sensitive cases of cervical
spondylitis the patient can eat with less motion and
less pain.

Traction can be used with advantage only in cases
confined to bed, being especially advantageous in the

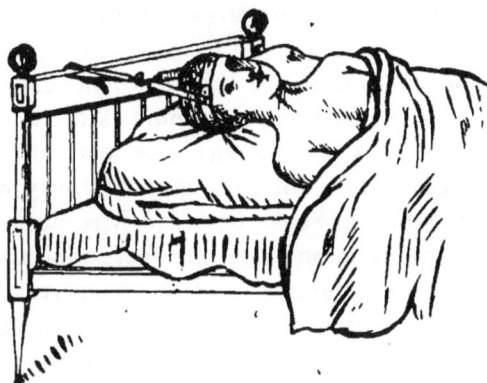

Fig. 42.—Elastic head-traction to bed-frame.

presence of cervical spondylitis, less so in lumbar and
dorsal disease. It is of value when inflammation is
progressive and when paraplegia complicates. , It is an
efficient aid in reducing to a minimum the pain and increase of deformity from muscular spasm during the

formation of an abscess. Traction does not appear to separate healthy articular surfaces or diseased ones after the reparative process has commenced. It does, however, at times reduce the deformity, apparently by separating the contiguous carious surfaces, and this without pain or any ascertainable untoward results.

In cases successfully treated by Mr. Jones the little patients are strapped on canvas stretched on a frame that rests on the bed on four short legs. Straps are placed

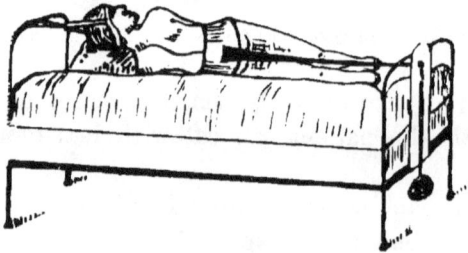

FIG. 43.—Fixed traction from head to bed-frame, with weight-and-pulley traction from pelvic girdle.

that fix the shoulders and thighs to the canvas, and holes opposite the anus and perineum assure easy egress to excretions. By this device the patients can be moved about without interfering in any way with the diseased vertebræ. Sayre uses the wire cuirass with head-sling and traction from the feet; this apparatus cannot be readily obtained and is expensive. Steele, of St. Louis, attains the same end by a portable stretcher-bed, consisting of a forged, oblong frame of flat bar-iron, made somewhat longer and slightly wider than the patient, over which are snugly stretched two pieces of strong canvas, one reaching from the buttocks to the top of the frame and the other from just below the buttocks to the bottom of the frame, the space between the two being left for the use of the bedpan. Upon this stretcher the patient is placed, strapped down

at the shoulders and hips, and traction is made from
the head-sling upward, to a flange at the top of
the frame, by means of an elastic or inelastic strap,
and downward by elastic or inelastic-straps attached to

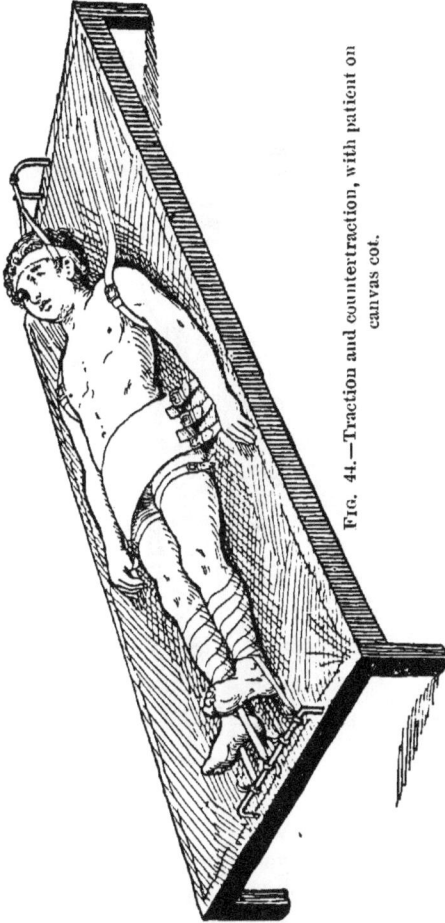

FIG. 44.—Traction and countertraction, with patient on canvas cot.

strips of adhesive plaster applied to the patient's legs,
to two flanges at the bottom of the frame. The patient thus lying on the frame can rest upon the bed or
be carried about without discomfort of motion to the

spine. The iron-work of the frame can readily be
done by any blacksmith, and the covering by the family of the patient; all should cost but very little.

The surgeons of the Children's Hospital in Boston
reduce the expense still further by making the frame
of iron gas-piping, and obtain traction from head and
legs by the ordinary weight and pulley to the head and

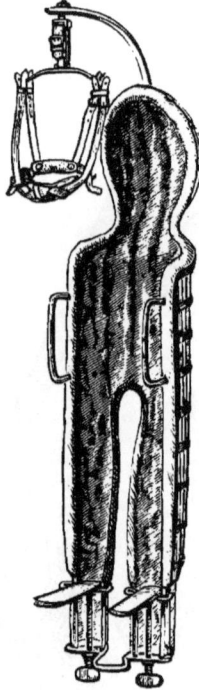

FIG. 45.—Sayre's cuirass.

foot of the bed. A gaspipe frame can be made to
take the place of Steele's frame by having pieces of
smaller pipe set upright in the middle of the bar at the
head and like pieces at each of the corners at the foot. An
excellent device is to slightly modify a Thomas double
hip-splint by putting the parallel bars a little closer to
each other and extra side wings so as to restrain all lat-

eral motion at the hips. A piece of strong leather can be
stretched from one main stem to the other to form a sling
for the spine and make it possible to lift and move the
patient as one piece. Without some such plan, to turn
a patient over for cleaning or other purposes, means

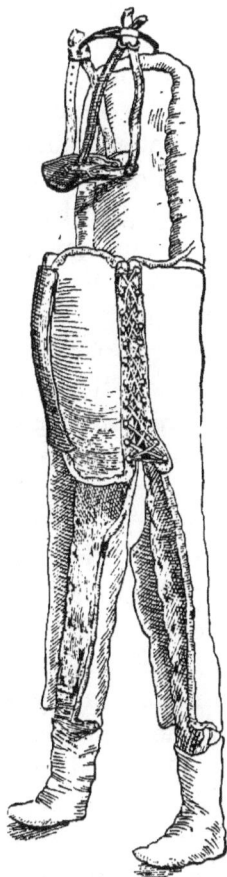

FIG. 46.—Phelps' plaster-bed.

damage, slight or severe, to the carious column.
Thomas' modification of the Bauer support offers
admirable assistance to bed-treatment if two bars be
added which extend to the knees and thus fix the
thighs. This appliance is easily borne and its applica-

tion offers no difficulty. The irregularities of the bed are thus obviated, and the greatest leverage can be employed to modify existing and prevent threatening deformity. Another device of simple construction is the Phelps plaster-bed. It is constructed in the following manner: A ¾-inch board is cut roughly to the form of the patient, with the legs somewhat separated; foot-pieces are put on at the bottom and an iron flange is set in at the top to carry the head-sling; the upper surface of the board is upholstered and the patient is laid thereon; then he is wrapped from the shoulders downward, together with the board, in plaster-bandages; before the plaster fully hardens the front is cut out, leaving a plaster-trough, the cut edges of which are bound, and in which the patient can comfortably lie.

Treatment with the Plaster-of-Paris Jacket.—This aims at immobilization after the patient has gained a position of greatest comfort by partial suspension. The opponents of the plaster-jacket treatment have asserted on the one hand that, by suspension, the carious surfaces are separated and the patient's life thereby endangered; and, on the other hand, that suspension is of no use inasmuch as it does not straighten the spine at the area of disease, but only apparently elongates it by straightening the normal curves. It appears to us that neither of these objections has any foundation in fact. There appears to be no evidence that separation of the carious surfaces by partial suspension has ever been of the slightest harm to the patient; and it has not been claimed by the advocates of suspension that it would straighten the curve of disease after reparative consolidation had at all advanced. Portions of curvatures and whole curvatures due to involuntary muscular spasm, and angles due to loss of bony tissue in the vertebral bodies can, before any considerable reparative action has taken place, be in a measure sometimes totally rectified by well-judged and carefully executed partial suspension.[2]

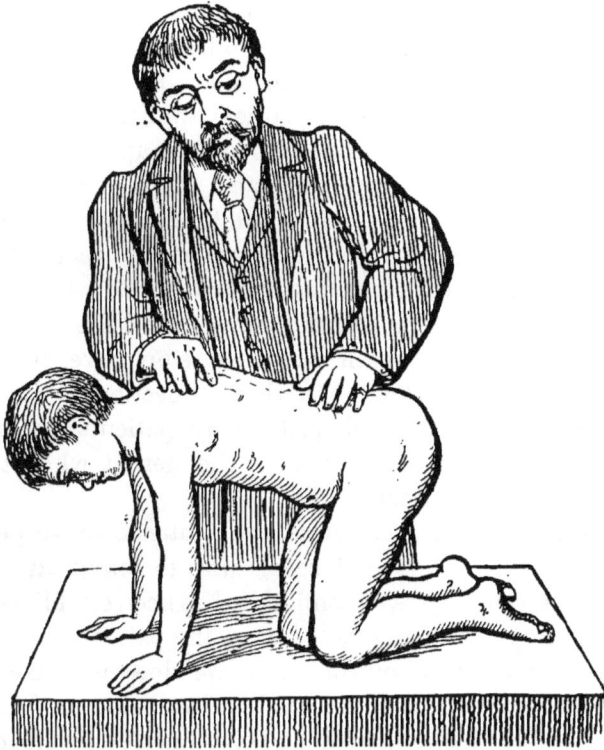

Fɪɢ. 47.—The hand-knee posture for applying a plaster-jacket.

The plaster-jacket can, of course, be applied without

---

[2] When the manuscript of the part of this chapter relating to the plaster-jacket was submitted to Dr. Sayre for suggestions and corrections he told Dr. Ridlon that a patient had been killed in Berlin by the breaking internally of an abscess-wall during complete suspension with weights attached to the feet and during chloroform-narcosis. Mr. Jones adds to this his experience in two cases, neither of which was published at the time. He was called to see a patient who had returned home from one of the hospitals after having been suspended during the application of a jacket for upper dorsal curvature. The patient, who had been perfectly well up to the period of suspension, died 2 hours after leaving the hospital, after complaining of pain in the limbs and suffering great respiratory difficulty. In the other case paraplegia resulted suddenly, being almost complete in 24 hours. In both cases suspension had been too complete, although in each case a surgeon of repute superintended the application. Dr. Sayre expresses a doubt as to the possibility of fully correcting any true angle by the suspension-treatment.

suspension of the patient, but unless the spine be put in a position of greatest comfort to the patient the object for which the plaster-jacket was designed is not attained, and failure to gain good results should not be accredited to the jacket-treatment. Surgeons are much too prone to modify methods and mechanical appliances without duly appreciating the principles of the apparatus that they ingeniously "improve" and label with their names. It is safe to say that of the thousands who have used the plaster-jacket in the treatment of spondylitis very few have ever given due thought to the teaching of Dr. Sayre to " suspend the patient until the point of entire freedom from pain is reached, stop there, and at once apply the jacket." In this connection, however, it must be remembered that many patients do not complain of pain, even during the period when the angle is on the increase.

Some very young children are frightened by suspension, and in their case it seems wise to forego its use until a certain degree of confidence has been established. In a few cases it is not well borne, the patient showing such a tendency to syncope that one does not like to repeat the experience. In such cases, prior to fixing with plaster, the hammock of Davy may be employed, or the patient may be placed in the hand-knee posture and the spine guided into the best possible position by gentle manipulation. If the hammock of Davy be used it must be drawn tight; otherwise the sagging as the patient lies with the face downward will give an uncomfortable position when the jacket has set and the patient stands.

The appliances requisite for suspension are as follows : A strong hook set into a beam, or a tripod, or crane. To the hook are attached a block and tackle, which support an iron cross-bar from 12 to 18 inches long, grooved transversely for adjusting the leather head-sling, or collar, and arm-slings that hang from it. The collar and arm-slings are so adjusted upon the patient that, when he stands directly under the cross-bar and traction is made upon the pulleys, the force is expended equally upon the head and arms.

For the jacket the plaster-bandages should be made by the surgeon or under his immediate supervision, for we know of no place where even fairly good ones can be purchased. Dr. Sayre has them made from

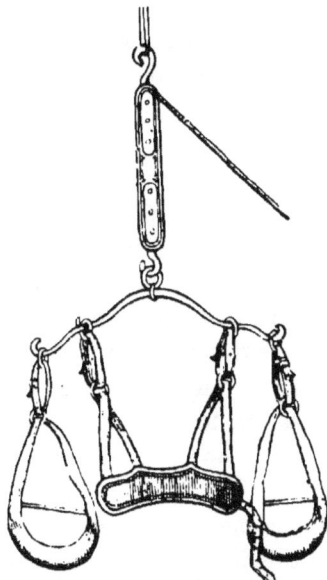

FIG. 48.—Suspension-apparatus.

cross-barred crinoline, in lengths of from 3 to 5 yards, torn into strips 2, 4, or 6 inches wide, according to the size of the patient, care being taken to tear off the selvage from the fabric.[3]

---

[3] The starch used in stiffening the crinoline in no way interferes with the setting of the plaster, but all specimens of cross-barred crinoline that we have found for many years past have been stiffened with some glutinous substance that delays the setting of the plaster. Such stiffening must be washed out and the goods ironed before being used. For some time past we have used a crinoline, not cross-barred, made especially for use in plaster-bandages by The H. B. Claflin Co., of New York, and known as "H" crinoline. It comes in pieces of 27 inches wide and 12 yards long, and costs 45 cents per piece. A piece is sufficient for 18 bandages, 3 inches wide and 6 yards long.

To make plaster-bandages the strips of crinoline are spread upon the flat surface of a table or shallow tray and the best quality of dental plaster-of-Paris is thoroughly rubbed in, removing the excess, and rolling the bandage rather loosely as the plaster is rubbed in. A bandage rolled tightly requires too long a time to become thoroughly wetted, while the center of one rolled loosely and with too much plaster between the rolls easily slips when wetted and becomes twisted and tangled. Too much plaster between the rolls of the bandage is a very common fault, and needs to be guarded against. It is our experience that bandages made in any other way and from any other material, however satisfactory for ordinary plaster-splints, will be found of little use for making really durable plaster-jackets. From 7 to 15 bandages will be required to make a jacket; if the jacket is to be cut down it should not be made too thin. For soaking the bandages a pail is used with sufficient tepid water to cover, by 2 or 3 inches, the widest bandage when standing on its end. It will not be necessary to add salt or alum to the water to hasten the setting; nor should the water be too hot. The plaster-sediment left in the bottom of the pail in which the bandages have been soaked, should not be used to rub into the jacket, as it will greatly delay the setting of the plaster and even soften that which is already set. A competent assistant is of the greatest importance, and he should rapidly and carefully smooth out every wrinkle and rub well in every layer as it is laid on.

The patient should be clothed in a seamless, skin-fitting knitted vest, made long enough to reach below the middle of the thighs, and well pulled down; it should fit without a wrinkle or a loose place. The surgical-instrument shops usually carry wool shirts of this order, but recently we have generally used tubular

stockinet of cotton of gray color. This stockinet can
be obtained of the knitting-factories in pieces of any

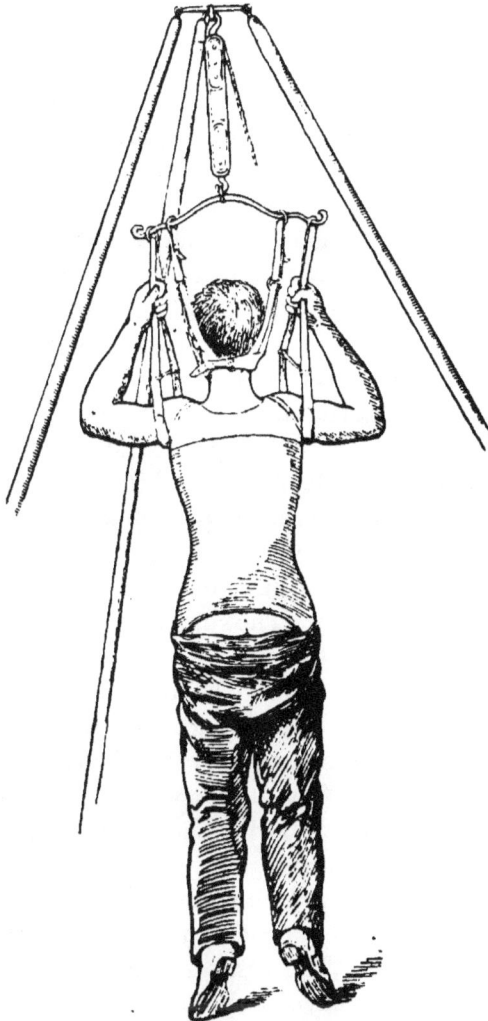

FIG. 49.—Plaster-jacket applied with the
patient partially suspended.

number of yards, and can be cut in any required length.
It is much less expensive than the wool vests. All other

clothing should be removed down to the level of the greater trochanters.

If the patient be a woman or an adolescent girl, breast-shields, or in lieu of these pads of cotton-wadding of proper size, should be placed between the breasts and the shirt; and if the jacket is to be made removable a strip of zinc or block-tin 2 inches wide and long enough to reach from the neck to the pubes should be placed under the shirt for protection to the patient when rapidly cutting off the jacket. On the outer side of the shirt pads of felt should be placed over the iliac crests and long, narrow strips along each side of the spinous processes included in the kyphosis. The floor should be covered with a sheet; and two chairs placed for the surgeon and his assistant.

Now, all being ready, the patient should stand beneath the suspension-apparatus while the surgeon adjusts the collar and arm-slings, and suspends him to the point of comfort, and no further. The assistant sits in front and grasps with his knees the thighs of the patient and steadies him. The surgeon sits behind, places a bandage on end in the pail of water and waits until the air-bubbles cease to rise; he then puts in a second bandage, squeezes out the superfluous water from the first, and rapidly and smoothly winds it around the patient's waist, and from there works downward over the iliac crests to a level with the greater trochanters; then he works upward again, each turn of bandage overlapping the former by two-thirds of its width, until a point is reached at the back and front well above the level of the axillæ. The assistant must smooth out all folds and rub each layer well into the preceding one. In this way the bandages are laid on until a sufficient thickness has been attained. Then with a well-sharpened knife the jacket is trimmed out under the arms and at the flexures of the thighs, so that the patient may afterwards sit with comfort.

By this time the jacket will usually have become
sufficiently hard to permit of a discontinuance of the
suspension. The patient should then sit quietly until
the setting of the plaster is quite complete. If the
plaster sets slowly, or if for any reason the time of
suspension has to be shortened, the surgeon, placing

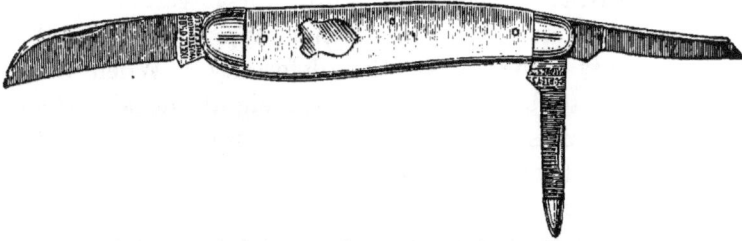

FIG. 50.—Ridlon's plaster-knife, made by Wostenholm, of Sheffield Eng., and
imported by J. Curley & Bro., 6 Warren St., New York City. This knife
is specially shaped and specially hardened for plaster-work.

his hands under the patient's arms, lifts him while the
assistant, after removing the collar and arm-slings, supports the patient by the thighs; thus he is placed prone
upon a couch to await the completion of the work. But
if the jacket is to be made removable it is cut down
from neck to pubes, while the patient is still suspended,
carefully sprung off the patient, its cut edges brought
accurately together and held by an ordinary gauze or
muslin roller-bandage. The jacket is then set aside to
dry—it will usually take 2 or 3 days—or it may be
rapidly dried in an oven or over a range, in which case
it must be carefully watched less it become brittle.
When dry it is tried on the patient during partial suspension, and trimmed wherever it may be necessary to
render the patient quite comfortable; the outer and
inner layers of the shirt are stitched together over the
cut edges in front on either side, and here on each side
on the outer surface of the jacket are sewed two strips
of strong leather, previously provided with lacing-hooks
set at intervals of about an inch. The patient is then

clothed in a well-fitting undervest—those made to measure and of Angora wool, and skin-fitting are the best—suspended as before, and the jacket applied and laced. The jacket must not be removed at night, or at any time except during partial suspension and in the presence of the surgeon.

For disease above the eighth dorsal vertebra, the jacket alone does not give sufficient support to prevent the steady progress of the deformity. When the disease exists between the first and eighth dorsal vertebra a jury-mast should be used to support the weight of the head, and more especially to prevent it from drooping forward.

The jury-mast consists of strips of tin, perforated in opposite directions, and joined to 2 steel uprights, at the back, bent to fit the outline of the patient. The tin strips, 2 on either side for a child and 3 for an adult, extend laterally from the posterior steel uprights nearly to the median line in front, but not across the spine at the back, and with the posterior steel bars are worked in between the layers of the jacket during the process of construction. The posterior bars are curved at the top, approach each other and are joined into one by being welded to form the upright bar that passes upward over the top of the head. This bar is bent to approximately follow the contour of the neck and head, and may be lapped and fastened with screws at the back of the neck so as to be elongated at will. It ends directly over the top of the head, and to its upper surface is riveted a cross-bar, turned up at the ends, from which depend the head-slings. The cross-bar being riveted by a single rivet, loosely set, the patient is able to turn his head from side to side at will.

For disease in the cervical vertebræ it is customary to make use of the same appliance, but we have not found it to immobilize effectively.

After the application of a permanent plaster-jacket no patient should ever pass from the surgeon's immediate control before 24 hours have passed in perfect comfort; any complaint of it hurting at any point, then or later, should be considered as a positive indication for the removal of the jacket.

The objections that may be urged against the plaster-jacket are chiefly its cost and the delusion that most surgeons labor under that it is a simple thing to properly apply it. It will be evident from the foregoing that it

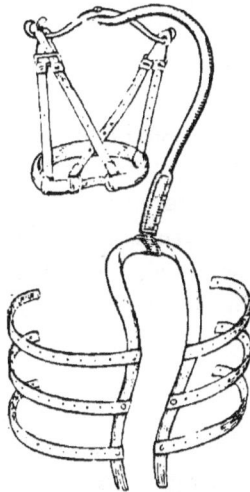

FIG. 51.—The jury-mast.

is not the ideal treatment for dispensary-work if little time can be devoted to each patient and if every detail of cost is counted. A really good jacket will last from 2 to 3 months if a laced one, or somewhat longer if permanent, but a growing child will require from 4 to 6 jackets each year and the disease will require treatment from 2 to 6 years. When to the cost of materials is added the value of the surgeon's time it will be found to be an expensive method of treatment.

Grave objections, however, may be urged against

plaster-jackets improperly applied, as it seems to us they usually are, from observations based upon patients wearing jackets applied at hospitals and dispensaries and by the family-doctor. With few exceptions they have been permanent jackets, seldom padded, and never with felt, over the bony prominences, rarely carried high enough or sufficiently low, generally so loose that the hand can be readily passed under them, so lacking in power to immobilize that the patients give a history of steady and progressive growth of the deformity, and when used among those of our fellow-citizens who have with reason been called the "great unwashed" have formed the pleasantest of homes for vermin of various sorts.

One does not of course have the opportunity to remove jackets applied by others from those patients who have done well and are satisfied with the treatment; but from the other patients, those that have not done well and are not satisfied, jackets are rarely removed without pressure-sores being exposed.

The Treatment of Spondylitis by the Antero-posterior Leverage Spinal Brace.—It is not necessary to describe the orignal Taylor brace, as it is no longer in use. The modifications of this brace by Dr. Taylor have all been designed to increase its efficiency; those introduced by others have generally been to reduce its cost, but the principles upon which they have all been used are the same, namely, immobilization of the spine at the area of disease by adjustable leverage, using the transverse processes of the vertebræ included in the kyphosis as the fulcrum. Braces have been made before and have been made since the Taylor brace was devised, much like it in appearance, but generally differing from it by separating the parallel upright bars so far that the leverage is brought to bear on the ribs, or they have attempted to combine traction with leverage and have thus failed to effectively apply the principle of making the leverage adjustable.

By adjustable leverage in the treatment of kyphosis, it is meant that the brace is so constructed that by manipulating the bars by means of wrenches (at first it was by hinges and set-screws) the pressure over the transverse processes of the vertebræ composing the angle can be adjusted to a nicety and increased or diminished at will, it being anticipated that in certain cases and with the disease in certain localities the curve will gradually diminish and occasionally be entirely effaced by the leverage-action.

It will be seen that the principle is essentially different from that underlying the use of the plaster-jacket. It does not suspend or partially suspend the patient, to gain the posture of greatest comfort or to improve the curve, but it applies the brace to the patient, with no attempt at improvement in his posture beyond that which is gained by lying down for a short time. More often than otherwise the patient is kept recumbent only so long as it takes to apply the brace, and, at times, when the disease is in the lumbar or in the cervical region, the brace is applied with the patient standing. This, however, is contrary to the teaching of Dr. Taylor, who never permits the patient to stand, either during the application of the brace or afterward, until convalescence is well established.

The brace being applied, the chief aim is to immobilize the area of disease until a cure is effected; meantime, if consolidation has not already taken place, it is attempted by gradually increasing the pressure to straighten the curvature, or at least check the progress of the deformity. The plaster-jacket aims at preserving the reduction of deformity gained by periodic partial suspension; the brace by its continued leverage-action is used to reduce the deformity; both primarily aim at immobilization.

The advantages of the leverage-brace, over and above the efficiency with which it carries out the principles involved, are its comparatively small cost, its durability, and the little time and effort required of the surgeon to adjust it. A gunsmith, locksmith, or blacksmith of average ability can, under surgical supervision, be trusted to make it, and the result will be a more efficient apparatus than can be obtained from any of the great instrument-shops, where exorbitant prices and erroneous ideas as to construction are usually prevalent. Braces with hard-rubber pad-plates and bearings will cost considerably more than if the rubber pieces be omitted, as the shaping of these hard-rubber pieces requires specially constructed molds and consumes much time. The brace, except when these pads are used, can generally be fitted in a few minutes; it does not require frequent modification when once properly fitted; and it rarely requires repair or material alteration; it is comparatively light, cool, and easily kept clean and free from vermin, and the patient can, while recumbent, have it removed without risk and enjoy the pleasure of a sponge-bath. To be sure, it requires a certain degree of mechanical knowledge to rapidly and perfectly adjust it, but it appears to us a less difficult proceeding than to properly apply a plaster-jacket.

The objections to the brace are that it is difficult to fit over a large kyphosis when the disease is in the upper dorsal and cervical regions, and even more so when any considerable lateral deviation exists; that prolonged use of the chin-piece may cause some recession of the chin; and that it makes no provision to prevent forward bending of the spine immediately above and below the area of disease.

This brace is not an apparatus that can be bought ready-made at any instrument-maker's; it must be made for the individual, from measurement and pattern, and with a definite end in view. A tracing is taken with a flexible metal tape along the line of the transverse processes—the line of the spinous processes always shows a much greater curve—and the tracing is copied upon strong paper, whereon are noted the position and direction of the shoulder-pieces, the place of the cross-pieces and the pad-plates, and the length and curve of the hip-piece. This diagram is the guide for the instrument-maker. The resulting brace should accurately follow it; but it usually requires a little refitting, which is accomplished by gradually bending it with wrenches made for the purpose; or ordinary monkey-wrenches may be used.

The pad-plates must be made to accurately fit the surface upon which they bear, and the remainder of the brace to approximately follow the outline of the back. The test of an accurate fit is a pink pressure-line upon the skin for the full length of the pad-plates after the brace has been worn, but with no place showing that the pressure is sufficient to create discomfort or to cause sloughing. The brace should be removed every day by the attendant while the patient is prone; the back should be washed, rubbed with equal parts of spirit and water, and when dry dusted with some good talcum-powder or a mixture of powdered zinc oxid and starch. The brace should not be removed at night until the patient is convalescent.

The form of brace now used by Dr. Henry Ling Taylor consists of two parallel upright bars, two shoulder-pieces, one cross-piece, a hip-piece, a chest-piece, an apron, and the connecting straps and buckles. (Fig. 52.) Each of the parallel upright bars consists of three parts: a forged pad-plate to which are attached an upper and a lower bar by a half-hinge, and a set-screw forming a false hinge. This false hinge is placed opposite the angle of deformity, and the pad-plates are made to extend well above and below the area of disease. The lower sections of the upright bars extend downward to a point just above the posterior spines of the ilium; the upper ones extend upward to the base of the neck, and in cervical disease to the upper part of the neck. The uprights are joined at the top by a short cross-piece; opposite the lower border of the axillæ is another cross-piece extending two-thirds across the back, and provided with buckles at the ends; at the bottom the uprights are attached to the hip-piece.

The hip-piece is forged from steel in the shape of an inverted U. It extends across the back above the posterior

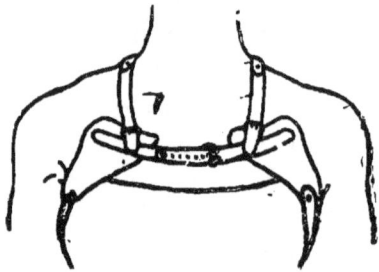

FIG. 53.—Showing ear-shaped chest-pads.

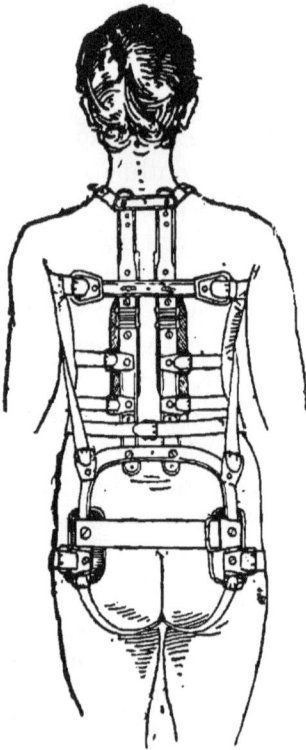

FIG. 52.—Dr. H. L. Taylor's support.

spines of the ilia and then curves downward to the hollow behind the greater trochanter on each side. At each end of the hip-piece is a bearing-pad of hard rubber, where a buckle faces downward and another laterally. Across the buttocks, at about the beginning of the anal fissure, is buckled a broad strip of webbing, passing from one side of the hip-piece to the other. Between the axillary cross-piece and the hip-piece, equidistant from these pieces and from each other, two buckles are attached to each upright, and face laterally. The bearing-surfaces of the pad-plates are lined with hard-rubber plates molded to fit the contour of the spine. The shoulder-pieces are of steel, attached to the uprights at such an angle that they may pass across the shoulders close to the root of the neck, terminating somewhat above the clavicles in straps that pass downward to buckle on the chest-piece. The chest-piece consists of two ear-shaped or somewhat triangular pieces of thin sheet-steel faced with hard rubber, shaped to fit the contour of the chest below the clavicles and in front of the shoulders, and joined by two steel bars lapped and screwed so as to be of adjustable length. (Fig. 53.) From the lower end of each ear-shaped piece a webbing-strap passes downward to a buckle at the bend of the hip-band. The apron that holds the whole apparatus in place is made of two thicknesses of twilled muslin, and reaches on each side from the axillæ to the iliac crests, and thence along the lines of the groins to the pubes, covering the entire front of the trunk as high as the arms. When the disease is situated at or above the seventh dorsal vertebra a head-piece is added, attached to the upper cross-piece by a pivot; this head-piece may be of the form shown in Fig. 58. The head-piece is an ovoid ring forged from steel; hinged on one side and fastened by a sliding-ring on the other. At the back a hole is made into which the pivot fits. Free lateral motion of the head-piece is permitted in dorsal disease, but in cervical disease this motion is restricted by a set-screw. In front, the chin rests in a hard-rubber cup, and at the back the occiput may be supported by two padded pieces of sheet-steel, screwed to the ring and extending upward.

An inexpensive modification of this brace is used at the Children's Hospital in Boston. (Figs. 54 and 55.)

The pad-plates are omitted and in place of the hard-rubber bearings stiff, smooth leather is used. The apron is narrowed at the top and made to extend upward to the sternoclavicular junction, to take the place of the chest-piece of the Taylor brace.

The form of antero-posterior leverage-brace that one of us (J. R.) has generally used, is shown in Figs. 56, 57, and 58. It consists of a hip-band, two parallel uprights, two cross-pieces, two shoulder-pieces, and two pad-plates. The hip-band is made of sheet-steel; it is

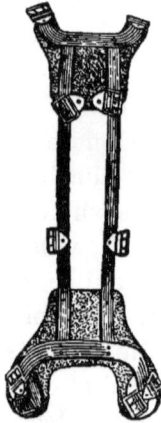 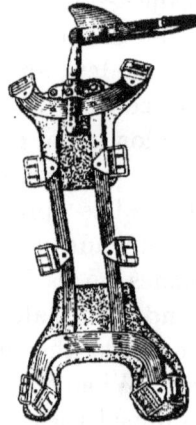

FIG. 54.   FIG. 55.

Braces used at the Children's Hospital, Boston.

from 1½ to 2 inches wide, and made of two pieces riveted together; the longer piece reaches from a point just above one great trochanter, across the back to a similar point on the opposite side; the shorter piece is one-third this length, lies upon the middle portion of the longer piece, and is riveted to it at the middle and at each end, before it has been bent, as later on it must be, to fit the outline of the hips. This arrangement gives a straight middle third more rigid than the same thickness in one piece would be, and an easily bent third at each end. At about an inch from each

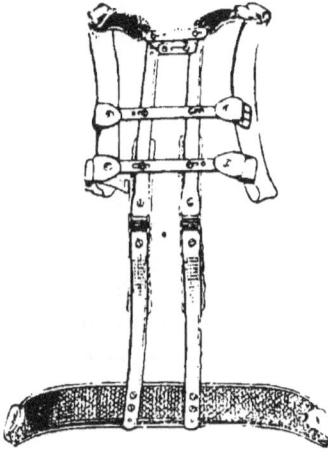

FIG. 56.

FIG. 57.

FIG. 58.

end a hole is bored for the attachment of a buckle; and at each side of the middle, three pairs of holes are bored for the attachment of the uprights. These holes are usually cut with screw-threads so that the uprights may be screwed on; they may, however, be riveted. The hip-band is lined on the side next the patient with felt, and the whole is covered

with leather. The uprights are made of annealed steel,
$\frac{1}{2}$, $\frac{9}{16}$, or $\frac{5}{8}$ inch wide, and gauge 9, 10, 11, or 12 in thickness, according to the size of the patient. They are
each pierced by three holes at the bottom, each hole
somewhat elongated and separate from the next by
the same distance as the holes in the hip-band. By
this arrangement the brace can be lengthened or shortened half an inch. The pad-plates may be simply
screwed on—holes having been pierced—or the arrangement of the pad-plates may be the same in detail as
that shown in Fig. 56. In any case, the holes through
the uprights, for screwing on the pad-plates, should be
elongated to allow for easy fitting after any change in
the curve of the brace. A hole is made in each upright at a point opposite the lower angle of the scapula,
for the attachment of the lower cross-bar, and another
pair of holes, opposite the lower borders of the axillæ
for the upper cross-bar. At the top of each upright
two or three holes are bored for the attachment of the
shoulder-pieces; if they are to be screwed on and made
adjustable, as is customary when no chin-piece is used,
screw-threads must be cut in these holes; when a chin-
piece is used, the shoulder-pieces are riveted on. The
pad-plates serve to strengthen the brace at the part of
greatest strain; if the false hinge is not required, they
are cut from sheet-steel the same width as the upright
bars and of a length sufficient to reach well above and
well below the kyphosis; they are pierced around the
border with numerous holes for sewing on the pads,
and, at about an inch from each end, a hole is bored
and cut with screw-threads for receiving the screws
that pass through the uprights. The cross-bars of steel
are somewhat thinner and narrower than the main uprights; in length, they extend for an inch or two to
each side of the uprights when in position. They are
pierced with a hole at each end for buckles, and with

three holes on one side and a slot on the other to allow of separation or approximation of the uprights. In putting on the buckles the rivet should pass from without inward, first through the leather and then through the steel, and be hammered into the hole in the crosspiece instead of being set into a bur.

FIG. 59.—Protective brace for convalescent cases. No apron in front.

The pads that are to be sewed on to the pad-plates are small bags of canton flannel, filled with powdered cork and quilted flat to about $\frac{3}{4}$ inch in width and $\frac{3}{8}$

inch in thickness. Good pads, however, may be cut from piano-felt. The shoulder-pieces are thinly padded on the side next the patient and covered with leather; at the end of each a tab of leather is riveted, and to these the shoulder-straps are sewed. The shoulder-straps may be of webbing covered with flannel, but they are better when made from a roll of blanketing or thin felt and covered canton flannel, and terminating in a short piece of webbing which buckles to the lower cross-piece. The apron is made of two thicknesses of twilled muslin, and reaches from the lower part of the abdomen to the level of the axillæ in front, and from the crests of the ilia to the axillæ laterally. Across the bottom is sewed a strong strap of webbing covered with canton flannel; this fastens to the buckles of the hip-band on each side. At each of the upper corners of the apron a piece of webbing is sewed in between the thicknesses of the muslin, and these pieces are fastened to the buckles at the ends of the upper cross-piece. Between the top and bottom straps, two or three others are sewed in along each side; all of these on one side are provided with buckles, to receive those from the other side when fastened around the patient and the brace.

When the disease is at or above the ninth dorsal vertebra, a head-support is used of the same form as that described in connection with the Taylor brace. With disease in the cervical spine a band may be riveted to the upper ends of the occipital rests and thence buckled around the forehead. When there is much rotary or lateral deformity in connection with cervical disease a ball-pivot may be used in place of the ordinary pivot, but this adds considerably to the expense, and it readily gets out of order. Under such conditions it is usually better to reduce the rotary or lateral deformity by horizontal traction and use the ordinary pivot, or in place

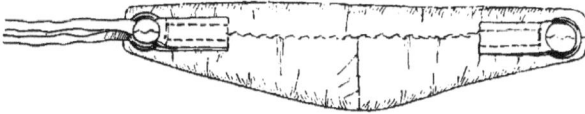

FIG. 60.—The Thomas collar.

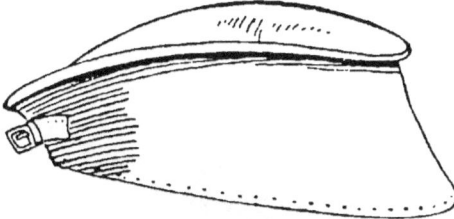

FIG. 61.—The Thomas collar.

of this brace the Thomas collar hereafter to be described.

To measure for a spine-brace of this kind it is necessary to transfer to strong paper a tracing of the spine taken with a lead or block-tin tape, and upon this should be marked the length of the hip-band, the location of the pad-plates, the cross-pieces and the shoulder-pieces. To an instrument-maker unfamiliar with the work the sizes of steel to be used should be specified and the pattern for the apron drawn.

The Thomas col-

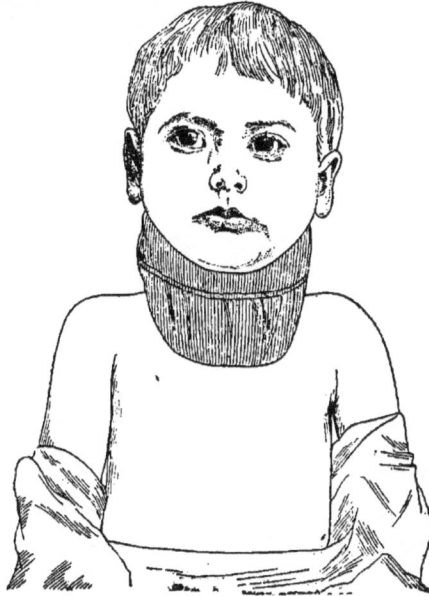

FIG. 62.—The Thomas collar applied.

lar, for use in disease of the cervical portion of the spine, is made by cutting from a piece of sheet-metal, steel, iron, aluminum, zinc, or tin, a piece straight on one side and convex on the other, long enough to somewhat more than encircle the neck; at the ends it

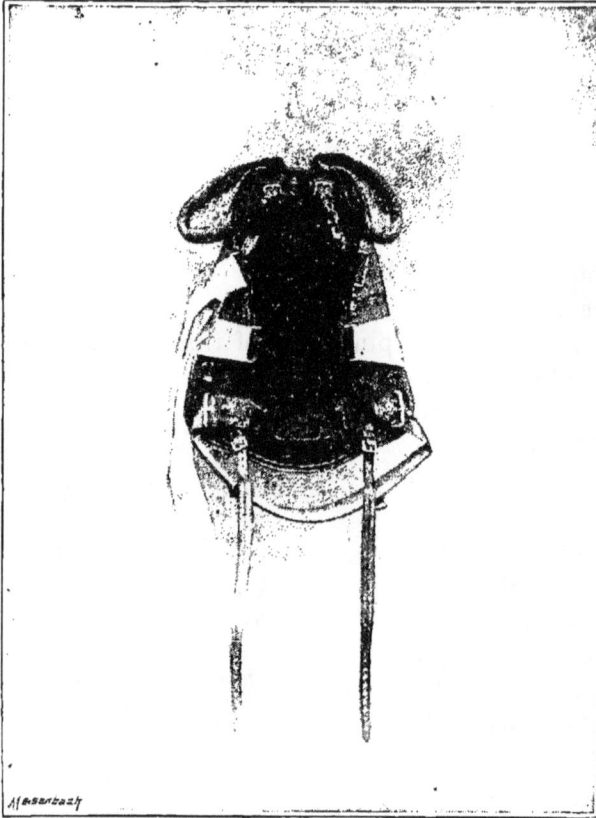

FIG. 63.—Thomas' cuirass.

should be wide enough to reach from the base of the neck to the base of the occiput, and in the middle wide enough to reach from the sternum to the chin. It is bent to roughly fit the neck; then the edges are turned slightly out, and the whole is wrapped in felt and covered with sheepskin. At the ends a buckle and strap are attached, or two rings, so that the collar may be securely fastened around the neck, resting on the chest and shoulders, and supporting the chin, jaw and occiput. This simple device is one of the most satisfactory of all methods for treating cervical caries.

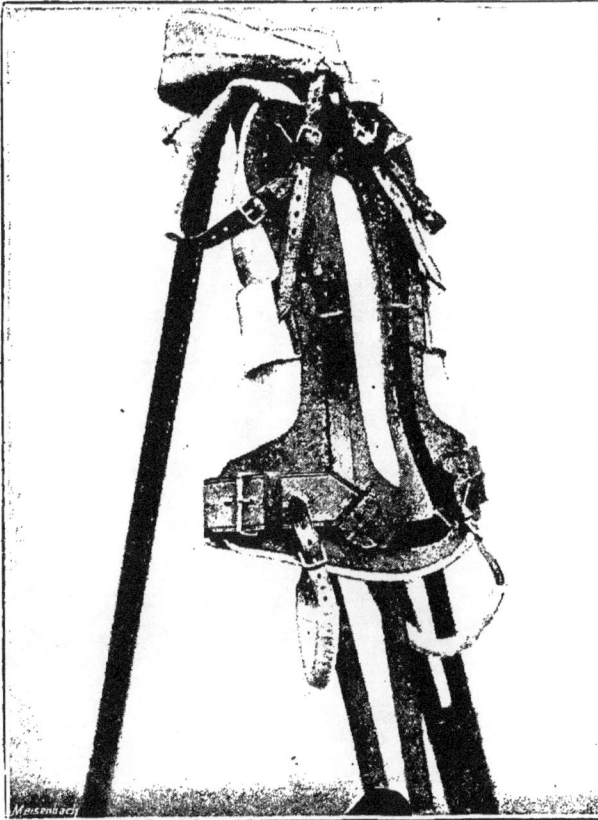

FIG. 64.—Showing cuirass before it is covered with leather.

Treatment of Spondylitis with the Thomas Cuirass.— The treatment of Pott's disease by this appliance (Figs. 63 to 67) is based upon the principle of immediate

and complete immobilization of the diseased area by
an apparatus applied in most cases to fit the deformity,
without any effort then, or at any time, to correct the
deformity by suspension, posture, and only very exceptionally by leverage. The principle logically obtains

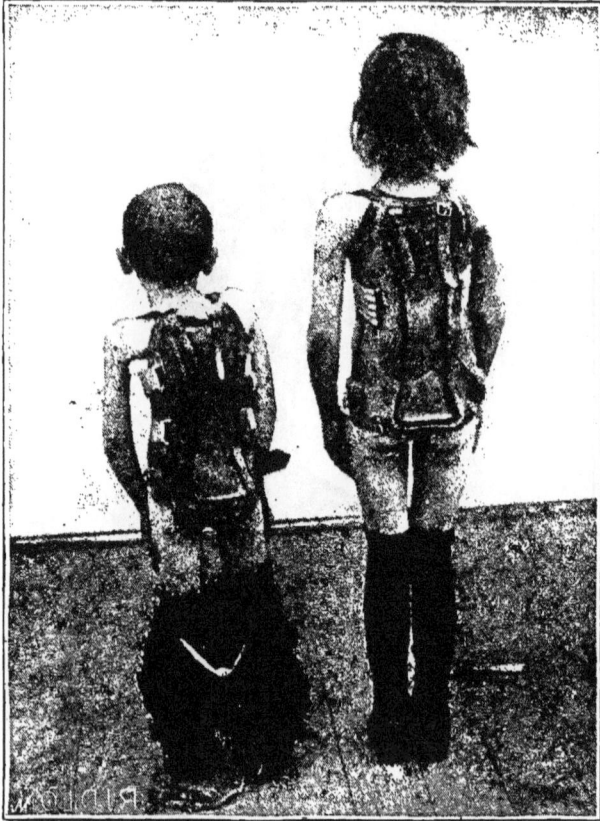

Fig. 65.—View of Thomas' cuirass applied to the patient.

from the theory that a diseased joint recovers quickest
when subjected to immediate and complete immobilization, and receives injury from, and is delayed in its
recovery by, each successful attempt at correction of
the existing deformity.

The brace consists of an irregular-shaped frame of flat bar-iron forged into the required form, as shown in Fig. 64. At the bottom it reaches to the level of the great trochanter; that is to say, it extends as low as the sitting posture will allow. Laterally it extends from the space posterior to the great trochanter on each side, and from there curves upward, passing to the outer side of the posterior superior spines of the ilium, thence inward to the immediate neighborhood of the spinal column in the dorso-lumbar region, from there curving somewhat outward toward the posterior border of the axillæ, then upward and inward to the back of the shoulders, at such distance as not to interfere with the movements of the arms, till the root of the neck is reached, when the two sides join in a horizontal upper bar. The width and thickness of the bar-iron used will depend upon the size and weight of the patient, but for a child of from 4 to 8 years it should be $\frac{3}{4}$ by $\frac{3}{16}$ inch. In forging the frame it is made to lie flat with some accuracy upon the patient's back. This frame, being in one continuous piece and nowhere pierced with holes, gives a great degree of rigidity for its weight. Under it is placed a piece of fairly rigid leather cut to the same shape as the frame, but extending beyond its margins as shown in Fig. 64. Again, under this is placed a sheet of saddler's felt extending a little beyond the borders of the leather piece. The felt and leather are sewed together, and to them are fastened the necessary straps and buckles. The whole is then covered with basil leather (Fig. 63). From the bottom of the brace a broad leather strap, lined with felt, buckles across the front of the patient, and secures the brace to the pelvis. At the lower lateral curves of the frame, on each side, a buckle faces downward to accommodate a perineal strap, which in front passes up to a buckle on the broad leather strap just mentioned. Above, at the junction of the neck and shoulder, a buckle looks forward and, at the lower

Fig. 66.—Front view of Thomas' cuirass.

border of the axilla, another looks laterally on either side; these are for the shoulder-straps. The shoulder and perineal straps are of felt covered with basil leather. From the middle of the brace on each side a strap of webbing two inches wide passes over the abdomen of the patient and buckles (Fig. 66). The position of this strap is changed with the necessities of the case, and at times a second strap is added.

Should the deformity be an extensive one and the angle formed by the spines of the diseased vertebræ be acute, one or both of two procedures may be necessary. The leather between the frame and over the kyphosis may have to be split, so that no pressure is exercised over sharp projecting bone; or, in addition, a bar of iron may be so placed over the projection as to render the recumbent position easy (Fig. 67).

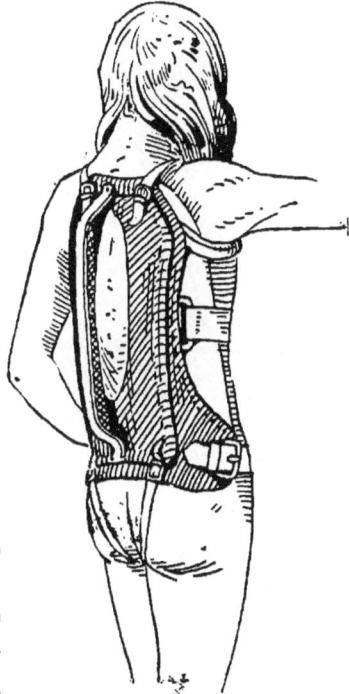

FIG. 67.—The Thomas cuirass with bridge for use during recumbency when the deformity is severe.

In exceptional cases, when the superincumbent spine falls considerably forward, traction is made by the shoulder-straps toward the cuirass, which, in such cases, in order to allow of a pull, is not fitted accurately to the upper portion of the back. In lumbar disease, or when there is psoas-contraction, a leg-piece is added, ending close above the knee, to prevent movement of the limb and traction upon the vertebræ.

One of us (R. J.) uses this support largely. It is

comparatively cheap and cleanly. It can be removed at intervals while the back is cleansed, and a sheet of cotton wadding inserted between the support and the skin. It need not be removed oftener than twice a week. This cleansing should always take place while the patient lies on his face, with arms outstretched above the head. The special value of this support con-

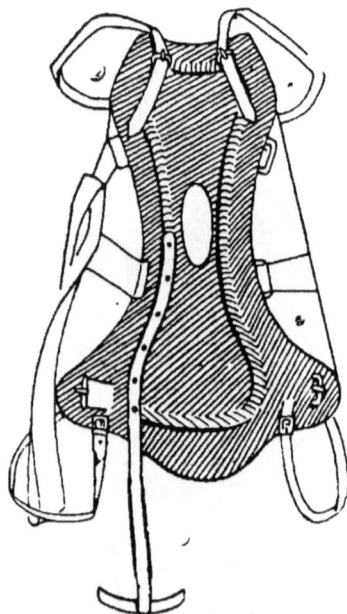

Fig. 68.—Thomas' cuirass with leg-attachment for reducing hip-deformity, and for treating hip-disease when co-existent with spine-disease.

sists in the length of the spine that it controls. It reaches the seventh cervical vertebra above, and by its action on the shoulders partly governs the upper dorsal vertebræ, while below it extends to the trochanteric regions and is there assisted by groin-straps. There is no undue pressure upon chest or muscles, and, with care, no danger of sores or excoriation. It is easily worn and is never uncomfortable and in no way interferes with recumbency. In order to measure for the s p l i n t the patient should be placed in a sitting position upon the chair and the distance measured from the seventh cervical vertebra to the chair. The measuring tape should not follow the contour of the back, but take the direction shown in the vertical dotted line (Fig. 69). In diseases high up, when the collar is required, it is well to cut out for the instrument-maker a pattern in brown paper something like the old-fashioned stock. Unless this be done it is very difficult to secure an accurate fit, as the position of the head and neck varies so much in different cases. When there is any doubt on the part of the surgeon as to his being able to measure properly for a collar it is well to order one filled with sawdust, which can be modified as to size and be molded so as to shape to suit the particular case.

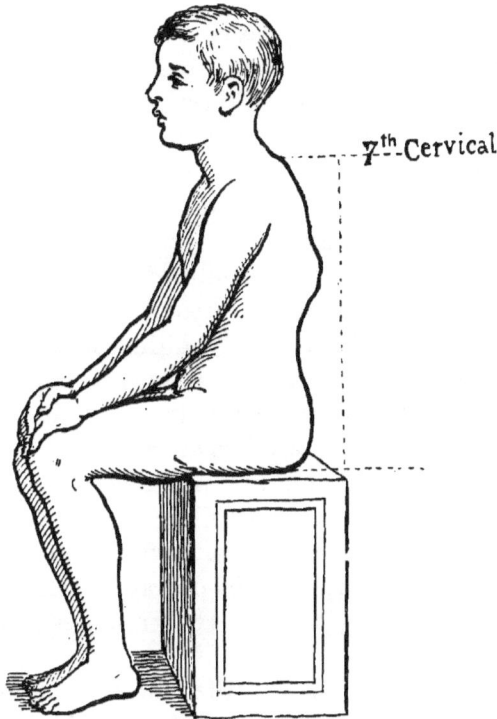

FIG. 69.—Method of measuring for the Thomas cuirass.

When the disease is in the upper dorsal region, a Thomas collar may be added and buckled (as shown in Fig. 64), or any of the head-rests and chin-pieces already described, may be attached to the upper portion of the frame.

The absence of holes, screws, and rivets renders the

construction of this brace simpler than that of the Taylor brace or any of its modifications, and while the patient is confined to recumbency it will be found more comfortable.

The operative measures for the treatment of spondylitis include aspiration of abscesses, with or without antiseptic injection; simple incision and drainage; and incision followed by erasion, with flushing with hot water, the wound being closed by suture and no drainage provided. Other operations attack the spinal column, either for the removal of necrosed bone or for erasion of the carious areas; and laminectomy may be performed for the relief of pressure-paralysis.

Simple aspiration, even when often repeated, has, as one might expect, not proved of much value. It rarely succeeds in completely withdrawing the abscess-contents in consequence of the caseous masses present, and is often followed by a rapid refilling of the sac. We sometimes see a single aspiration succeed in obliterating the abscess; but, more generally, even after repeated aspirations, it ultimately proves a failure. We often notice after aspiration a disappearance, sometimes lasting for many weeks, of all swelling, and then slowly the cavity refills. This points an obvious moral to those who publish cases as cured by operation before allowing sufficient time to elapse to render a fresh collection next to impossible.

Aspiration, followed by the injection of antiseptic fluids, has been abandoned, as also has the injection of iodoform-ether; but the injection of iodoform-"emulsion" (10% mixture of iodoform in olive-oil or glycerin) still finds favor with some general surgeons, and, in their hands, is reported to yield good results. The results, however, are less favorable with the large, tortuous and deeply-seated abscesses of spondylitis than with those at other joints. We have long since abandoned the use of this method of treatment, as always useless and at times harmful.

Incision of an abscess is demanded when the patient suffers from septic symptoms, when the location of the abscess is such as to prevent effective mechanical restraint to the diseased area, and when the abscess threatens important structures. Incision for the relief of such an abscess may have to be made almost anywhere. In cervical disease, behind the sterno-mastoid muscle; in dorsal disease, by the side of the vertebræ;

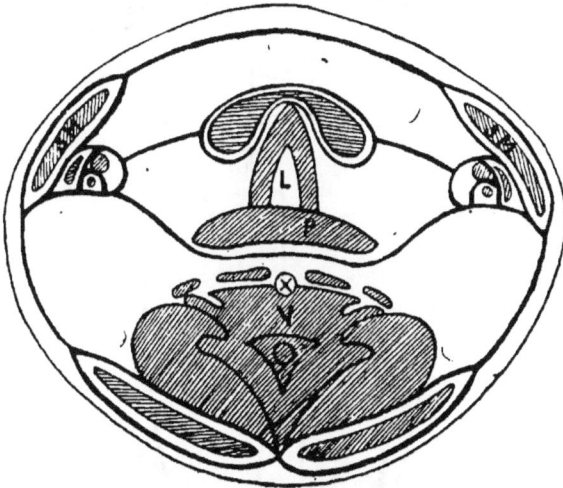

Fig. 70.—Diagrammatic section through the middle of the neck, showing the attachment of the pre-vertebral fascia laterally to the carotid sheath, thus directing pus into the posterior triangle of the neck in cervical caries. The fascia is attached above to the base of the skull; below it becomes lost in the posterior mediastinum centrally, and passing over the brachial plexus at the root of the neck in front of the subclavian artery to be attached to the costocoracoid membrane. If pus descends, it may find its way into the axilla or the posterior mediastinum, in addition to pointing in the pharynx and the posterior triangle of the neck.

in lumbar disease, just outside the erector spinæ, and in the case of psoas-abscess resulting from disease in any part of the spinal column, an opening may have to be made above or below Poupart's ligament, or even attack made from the lumbar region. The accompanying diagram shows the arrangement of the pre-vertebral fascia in the neck, and explains its attachments, which

have everything to do with the direction of pus in the
cervical region. The incision for the relief of abscess
in this neighborhood should be made at the posterior border of the sternomastoid, care being taken
to avoid the division of the spinal accessory nerve
(which leaves the sternomastoid at its middle). The
structures are then drawn forward with a broad retrac-
tor, when the transverse processes of the vertebræ will

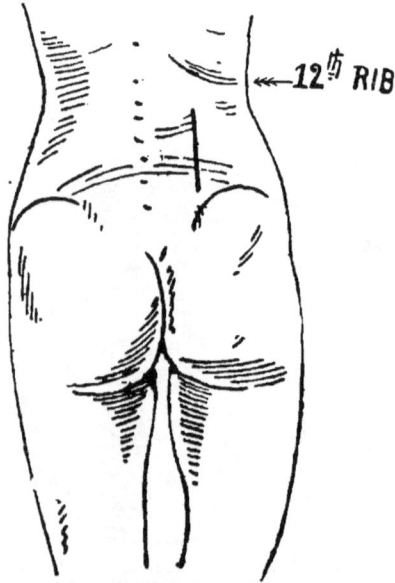

FIG. 71.—The lumbar incision.

be easily felt. If the abscess is large, it will be seen
bulging and can be easily opened (preferably by Hilton's method). The finger is introduced and diseased
bone felt for, and, if loose or easily detachable, removed.
This operation should be performed, however advanced
a retro-pharyngeal abscess may appear. It is a mistake
to incise through the pharynx, as drainage in any position of the body is thereby rendered difficult and
risky.

Dorsal abscesses may be opened where they present.

In the lumbar region, the vertebræ can easily be reached by an incision along the outer border of the erector spinæ, cutting through the posterior fasciæ and origin of the transversalis. The middle sheath and origin are now divided, and the quadratus lumborum is exposed. This muscle is easily known by the direction of its fibers, which pass upward and inward. The lumbar arteries cause no trouble, and the incision is kept as near to the middle line as the wound permits. By careful dissection arteries may be seen if present and they can be clamped before division. The anterior

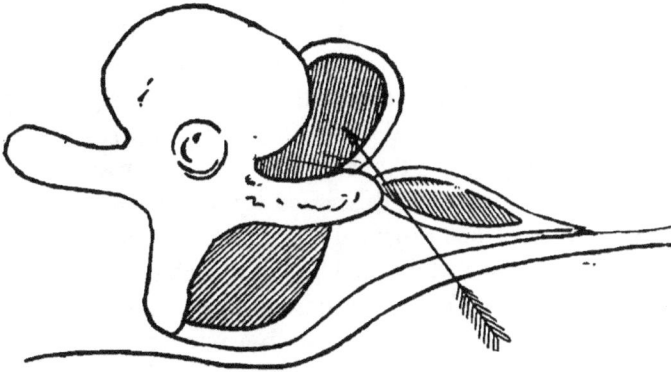

Fɪɢ. 72.—Direction of the lumbar incision.

fascia of the transversalis origin is now seen, then divided, and the finger passed along the front by the transverse processes and toward the bodies. The abscess can be easily felt by the finger, and fluctuation made out by pressing on the abscess in the thigh or abdomen from the front. The sides of the bodies can be made bare by a blunt dissecting tool or closed dissecting-forceps, and the exploration completed when sequestra may be removed. Thorough washing should be done. The abscess should be opened, and the cavity washed out, and scraped gently with the finger, or finger covered with gauze, when a slight general oozing of blood takes place but soon stops. The abscess should next be washed out with hot aseptic solution, preferably of boric acid. It will be found that after the cleansing the chronic abscess-wall collapses much more readily. The edges of the abscess-wall may then either be stitched or not. Deep sutures should be used for the quadratus and erector spinæ, and the superficial wound should then be closed. If this has been done carefully and thoroughly aseptically the whole cavity becomes obliterated by organization of blood-clot, which may fill the original abscess-cavity, now much collapsed. Before stitching has commenced, firm pressure is applied by an assistant pressing forcibly with both fists, one in the groin, the other over the abdominal wall, corresponding to the psoas sheath. This pressure must be continued until a firm pad is placed where each fist pressed tightly. The pads are then fastened by a carefully applied spica-bandage covering well over the abdominal wall. The loin-operation was first described by Treves.

A. E. Barker has reported cases in which he has successfully incised and flushed psoas-abscesses. He takes a case in which he presumes the bone-lesion to be stationary or healing, but in which a purulent collection is gathering. He makes a 2-inch incision through sound structures in the most dependent part of the swelling, after which a hollow gouge is inserted through the opening, and connected by piping with a reservoir of hot water at 105° to 110°. This reservoir (a three-gallon can) is raised up to 5 feet above the operating-table. The fundus of the abscess-cavity is by this means flushed, and the contents are washed away. The more solid caseous mass is dislodged by gently scraping with a scoop, until the soft lining membrane of the abscess is washed away. When the water runs out clear, the instrument is withdrawn, and all the water squeezed out. Iodoform-emulsion is then injected into the cavity, and stitches applied through the skin, the surplus iodoform-emulsion being squeezed out before the stitches are knotted ; the cavity is then closed with-out drainage.

Laminectomy for the relief of pressure-paralysis has been advocated by Macewen, Horsley, Lane, Willard, White, Lloyd, and others.

The patient lies in a prone position, and a pillow is placed under the lower ribs to produce a curve in the vertebral column, and an incision is made over the prominent spine long enough to admit of the free ex-posure of the laminæ by retractors, when the erector spinæ is cleaned from them. Transverse notches in the muscle will facilitate this and do no permanent injury, owing to the ankylosis of the vertebræ; owing to the curve in the spine this muscle is often easily drawn aside. The laminæ may be carefully sawed with a spinal saw, or cutters used specially for the pur-pose. The dura mater and cord are drawn to one side, and the tuberculous material at the back of the body gently scraped away.

The results of this operation are not such as to encourage its employment in any but the most desperate cases. It has distinct dangers of its own in its imme-diate effect upon the patient, and deprives the spinal column of practically its only support, when the bodies are largely eaten away by disease. It certainly should never be employed when thorough and prolonged mechanical treatment has not been tried. It is ex-tremely rare to find Pott's paraplegia permanent, and from an experience that has been exceptionally large, we can recall only two or three such cases, although we have experience of many in which the paralysis has lasted considerably over a year, and in a few for several years. With the recently revived operation of forcibly straightening carious spines additional hope is held out in these cases, and no case should be subjected to a cutting operation until forcible straightening has been tried and proved a failure.

DISEASE at the sacro-iliac articulation is of comparatively rare occurrence. Existing apart from spondylitis in the lower lumbar spine, it is of still rarer occurrence, and the diagnosis is so obscure that there are surgeons, careful observers and of extended experience in joint-diseases, who affirm that they have never met with it. For the most part, and perhaps always, the disease is tuberculous, and is governed by the same laws of pathology, symptomatology, and treatment that govern articular tuberculosis elsewhere.

Traumatism frequently appears to be the exciting cause, especially when the disease is found in young adults, but there can be no question that the disease occurs without any remembered injury, especially in those predisposed by heredity to tuberculous infection and rendered susceptible by debilitating diseases and the infectious diseases of childhood.

The disease may commence in either of the bones that go to form the joint, or in the synovial tissue within the joint. The bones more frequently appear to be the seat of the infection than the synovial tissue —there being no true synovial sac at this joint, but on account of the peculiar relations of the bones and because of the strength and thickness of the posterior ligaments and the absence of definite subjective symptoms in an early case, the disease is rarely recognized before suppuration has occurred, and all of the structures of the joint are involved. The disease may be of the so-called moist form and show early suppuration; or of the dry form, and run its course without suppuration; or the dry form under certain circumstances may at any time become suppurative.

Van Hook, who has made a most careful study of the literature of the subject, believes that the dry, non-suppurative form rarely imperils life and that the prognosis is in every way good, but that in the suppurative form the prognosis is exceptionally bad. It appears to us, however, that the symptoms detailed of many of the non - suppurative cases hardly warrant the diagnosis of sacro-iliac tuberculosis, and by that much detract from the weight that they would otherwise give to a favorable prognosis; and that the fatal termination and consequently unfavorable prognosis of the suppurative cases have more frequently been due to the character of the operative interference than to the nature of the affection.

There seems to us to be no good reason for believing that tuberculosis of the sacro-iliac articulation is governed in its fatalities by other laws

Fig. 73.—Characteristic attitude of sacro-iliac disease. From a photograph loaned by Dr. S. L. McCurdy.

than those governing the fatalities of tuberculosis of other joints, whilst our limited clinical experience of the disease goes to confirm this view. As in spondylitis, deaths occur from tuberculous infection of other organs quite as frequently in the dry as in the moist form of the disease, provided there be no operative interference. Death from prolonged suppuration is exceedingly rare when tuberculous abscesses are subjected to the let-alone treatment, and rarer still is death from septic infection. On the other hand there can be no reasonable doubt that any operative interference increases the risk of general tuberculous infection; and, unless the operation be strictly aseptic, and the prolonged subsequent dressings be kept so, the risk from septic infection of a large cavity connected with carious bone is considerable. In a word, any operation that fails to remove all tuberculous material and to close the cavity by primary union without drainage, though demanded as a last resort, should be recognized as distinctly adding to the risks of the patient's life. The records of the cases observed show that fatal termination is usually due to septicemia, simultaneous or intercurrent tuberculosis elsewhere, or general miliary tuberculosis.

The first symptom to appear is usually a peculiar attitude, a "listing" of the trunk toward the unaffected side, or, more properly speaking, a shifting of the hips toward the affected side; and as this progresses the spine assumes a long, sweeping curve, with the convexity to the sound side. Before the peculiar attitude has become sufficiently marked to cause comment, the patient usually finds himself fatigued on comparatively slight exertion, and has experienced difficulty in bending forward and rising up again. Ultimately, stooping is quite impossible. The gait becomes of a shuffling character, and as the disease advances the patient usually is unable to walk at all. In the early stage there is generally no flexion of the thigh, but, later on, this frequently appears and, with some degree of abduction or of adduction of the limb, simulates hip-disease or psoas-contraction of lumbar spondylitis. The abduction causes an apparent lengthening of the limb; the adduction an apparent shortening. The patient, standing, rests the heel upon the floor, but places nearly all his weight upon the sound leg.

The distant or referred pain, characteristic of tuberculous arthritis elsewhere, is usually present here, but may be absent. It is more frequently found in this affection than in disease of the hip or spine; if present, it is usually felt in the lower abdomen, but may be complained of anywhere along the front of the thigh and also along the area of distribution of the sciatic nerves.

At first the swelling of the joint-structures is more easily made out by palpation through the rectum, probably owing to the anterior sacro-iliac ligament offering much less resistance than the powerful and thick posterior ligament, and the swelling, therefore, is

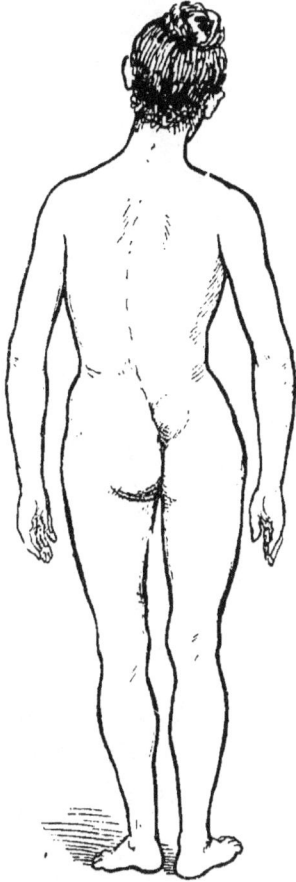

FIG. 74.—Right sacro-iliac disease. Leg abducted and gluteo-femoral crease lowered and nearly effaced.

directed toward the interior of the pelvis. Sooner or later the external swelling appears and in most cases advances to true fluctuation, and the tuberculous abscess is present as a complication. Such an abscess may extend in any direction; upward in the multifidus spinæ, into the lumbar region, downward along the psoas muscle, or into the buttock, to the right or to the left, or directly inward, to open into the bowel.

The direction in which the pus travels may be: (1) Through the anterior ligament, keeping outside the pelvic fascia; (*a*) following the course of the sacral nerves and pyriformis out through the great sacro-sciatic foramen and forming an abscess under the gluteus maximus; (*b*) following the curve of the sacrum behind the rectum to point in the ischio-rectal fossa, causing inflammation and adhesion of the rectum and ultimately bursting into it; (*c*) coursing under the lumbo-sacral ligament into the psoas muscle and thence to the thigh; (*d*) or into the iliacus muscle and thence into the groin. (2) Through the back part of the joint into the multifidus spinæ, creeping along this and pointing in the lumbar region or directly over the joint itself.

Muscular atrophy of the muscles of the buttock and

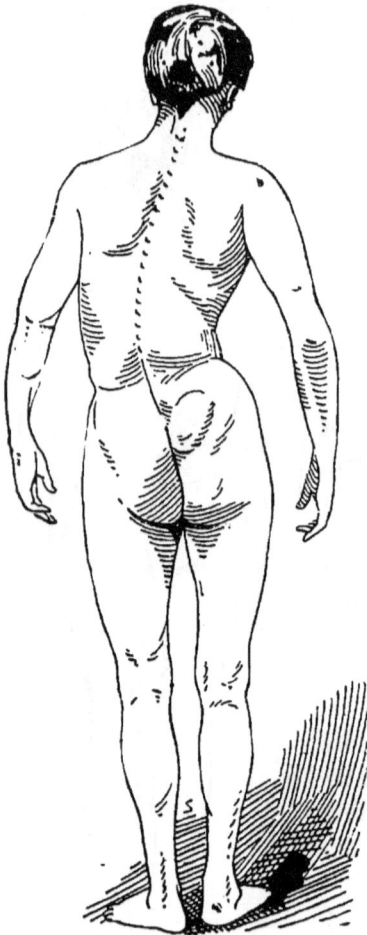

FIG. 75.—Right sacro-iliac disease. Leg adducted and gluteo-femoral crease raised.

thigh is uniformly present. Deep pressure over the articulation often causes pain before much, if any, swelling is noticeable, and pressing together or pulling apart the pelvic bones also induces pain. This pain appears to be due more to the motion imparted than to the

FIG. 76.—Abscess in right sacro-iliac disease.

direct pressure exerted. At times there is a tilting of the bones one upon the other, so as to form an oblique kyphosis, or a depression; or one bone may be elevated and the other depressed. Spasmodic contraction of the psoas muscle is a pretty constant symptom and may be found early in the disease; in consequence of this the thigh becomes somewhat flexed on the pelvis and rotated outward; hence the frequent confusion with hip-disease. Motion at the hip-joint may appear to be restricted in all directions, but if the pelvis be steadied and the manipulations be conducted with such gentleness as not to disturb the sacro-iliac joint, it will be found that, when the thigh is slightly flexed to relax the tension upon the psoas, all the hip-joint motions are normal except those that put the psoas on the stretch, namely, extension and inward rotation during extension. In the same way the contracted psoas muscle restricts the bending in the lumbar spine, and the resulting condition simulates lumbar spondylitis. Passive bending of the spine toward the affected articulation or forward when the patient is recumbent, if practised with great gentleness and with the pelvis steadied, will, by the freedom of movement, exclude spondylitis from the diagnosis.

The differential diagnosis is, for the most part, made from hip-disease and from spondylitis, and it can only be made by remembering that disease in any joint restricts not some but all of its normal movements to some extent. In some cases of sacro-iliac disease in which the muscular spasm and pain are intense, it may not at once be possible to differentiate, especially as the disease has been seen coincidently with hip-disease, and as it is more frequently found in conjunction with lumbar spondylitis than existing alone. The condition may be mistaken for sciatica, or for intrapelvic inflammation, or for abscess in connection with old recurrent appendicitis, but a careful examination and a consideration of the history of the case should clear up these points.

The mechanical treatment of sacro-iliac disease is not one of the most encouraging of orthopedic problems. It consists in a more or less successful attempt

at immobilization, but it is found far less easy to immobilize this joint than the hip or the spine, and satisfactory immobilization by an ambulatory apparatus is
practically out of the question. The ambulatory appara-
tuses that have seemed most successful have aimed at the
accomplishment of two
things: Immobilization
by circumferential com-
pression by a broad girdle
and limitation to volun-
tary motion by a spinal
apparatus that restricts
forward bending. There
is no question that motion
in the lumbar spine is
contraindicated and there
should also be no ques-
tion that motion at the
hip-joint is also contra-
indicated, but restriction
of the latter has not been
attempted by an ambu-
latory apparatus, as it
would prevent the pa-
tient from sitting. The
fact that the girdle, in a
certain number of cases,
relieves pain, which is
not relieved, but too often
aggravated, by traction,
points very suggestively
in the direction of the
true and of the false
principles of the treat-

Fig. 77.—Patient in apparatus. Brace
should be bandaged to the thighs just
above the knees.

ment of all joint-disease, namely, that a force that
tends to immobilize, even when associated with a force
that crowds together the articular surfaces, relieves

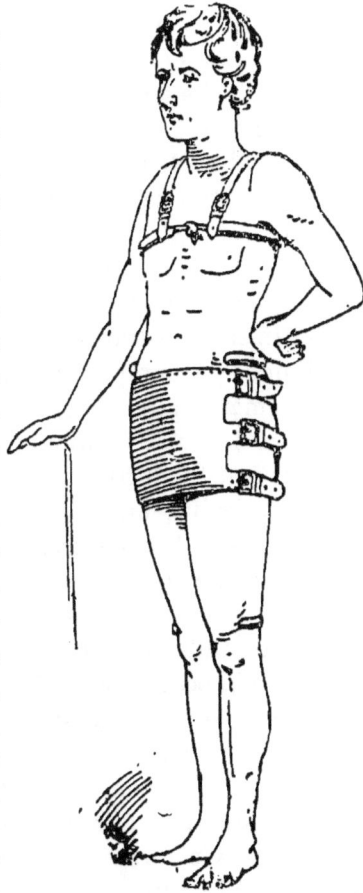

pain, whereas a force that tends to separate the joint-
surfaces without immobilization fails to relieve, and
often increases the suffering.

The mechanical treatment that we employ is some
one of the modifications of the Thomas double hip-
splint. The main stems should be separated at such a
distance as to pass to the outer side of the posterior
superior spines of the ilia, and lateral wings should be
attached to the stems to pass around the flank on either
side. One of us (R. J.) is accustomed to use the form
of splint depicted in Figs. 76 and 77, and the other
(J. R.) uses the same frame, but adds a broad sling of
leather stretched from one stem to the other and reaching from the coccyx to the mid-lumbar region; both of
us bandage the thighs to the stems. This is not depicted in the illustrations. The patient is to be kept
continuously recumbent for so long as any active symptoms remain; after that he may be allowed to stand
and to make such attempts at walking as his brace will
permit; but the hips must not be released, so that sitting is possible, until the surgeon is reasonably sure
that he has fully recovered.

Inasmuch as this disease usually appears in adult life
and but rarely in children, and inasmuch as the joint is
fairly accessible, we are of the opinion that as soon as
suppuration occurs, operative measures looking to the
removal of all tuberculous matter are to be considered,
and that such measures are justifiable in a larger per-
centage of cases than when disease attacks any of the
other joints. It is of advantage to prevent, when possible, intrapelvic burrowing, and this can be done without
our having to reflect, as we are forced to in the case of
the hip or the knee, upon an ankylosis that would be
harmless, or a shortening of limb that cannot occur.

The operative procedures are determined by the
facts learned from palpation externally and by way of
the rectum. If an abscess can be detected within the

pelvis, the incision is made directly down upon the ilium externally to this point, the bone trephined, the abscess-cavity gently and thoroughly cleansed, more bone removed, if necessary, with cutting-forceps or chisel, all cut bone-surface thoroughly seared with the actual cautery and the wound closed. If no point of fluctuation can be made out, the incision is determined by the edema, or, in the absence of edema, by the tender point. The bone is trephined for a caseating center, and the subsequent steps of the operation are as have been indicated. Recently Dr. L. L. McArthur has recommended an operation over the main articular portion of the joint in all instances as being the region likely to be chiefly diseased. The posterior inferior spine of the ilium and the sciatic notch are the guides, and that part of the ilium which goes to make up the anterior inferior portion of the joint is removed by saw and chisel. After any operative procedure the joint should be immobilized in the Thomas double hip-splint, and the patient confined

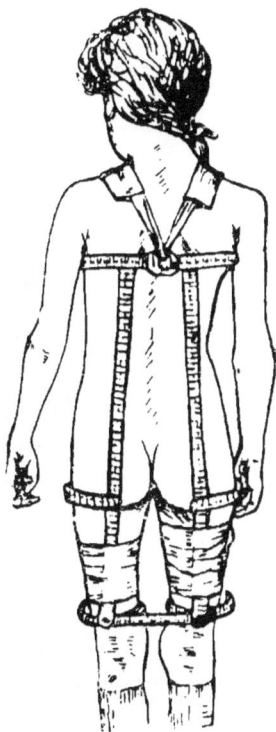

Fig. 78.—Apparatus for sacro-iliac disease. A leather sling may be stretched between the bars from waist to hips.

to bed until all local tenderness has passed away.

It is possible that there are more reasons to justify the use of the drainage-tube after operations upon this joint than upon others, but we believe that a second or several repetitions of the operation entail less risk than the insertion of the tube.

HIP-DISEASE.

The term hip-disease is used for any chronic inflammation of the synovial membrane of the hip-joint, of the acetabulum, of the head, neck, or greater trochanter of the femur, or of the soft parts immediately surrounding these, which, if allowed to progress without treatment, would ultimately present the symptoms of a tuberculous arthritis.

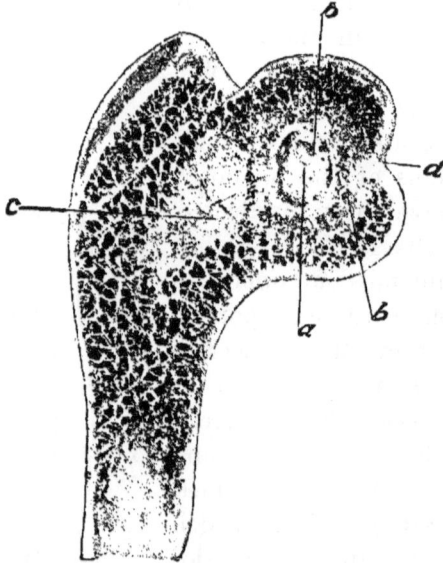

FIG. 79.—From a case of severe coxitis in a boy six years old. Primary tuberculous infection in the head of the femur. (*a*) Cheesy spot; (*bb*) infection of surrounding medulla; (*c*) extension into the shaft; (*d*) compression-groove where the head of the femur rested on the acetabulum. (Cut taken from Krause.)

The synonyms in common use are hip-joint disease, tuberculous arthritis of the hip, morbus coxae, and coxalgia.

The causes of hip-disease are tuberculosis, usually acquired, but occasionally inherited, inherited syphilis,

injury from falls, blows, and probably from jumping
and from sprains, the infectious diseases of childhood,
and the various causes that tend to chronic inflammation in and about bones and joints. But whatever the
cause, it ultimately presents the symptoms characteristic
of a tuberculous arthritis, and for all practical purposes
may be regarded as such.

The precise location of the pathologic lesion, as a rule,
can not be determined. From our knowledge of like
pathologic processes in more superficial joints we are
justified in assuming that the disease may begin in the
synovial membrane, but the symptoms of a chronic
synovitis at the hip-joint are very obscure, and a diag-

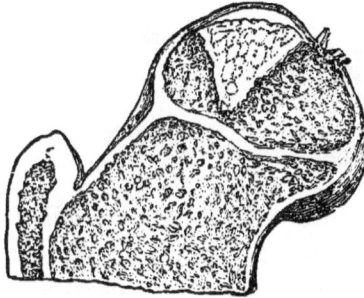

FIG. 80.—Resected upper end of a femur of a girl five years old. Large cone-
shaped subchondral focus with demarcation far advanced; articular cartilage lifted up like a vesicle. This is unquestionably a secondary infarction
focus. (Cut taken from Krause.)

nosis is rarely made before the bone is invaded through
the early destruction of the cotyloid ligament. When
the disease begins as an osteitis it is not possible to say
whether the primary focus is in the head, neck, or
greater trochanter of the femur, or in the acetabulum;
and in many instances it is not possible to say whether
the disease is articular or periarticular. This uncertainty as to the precise location of the primary lesion
renders early operative treatment ridiculous; without
any treatment all of the tissues of the joint ultimately
become involved; treated mechanically, many cases

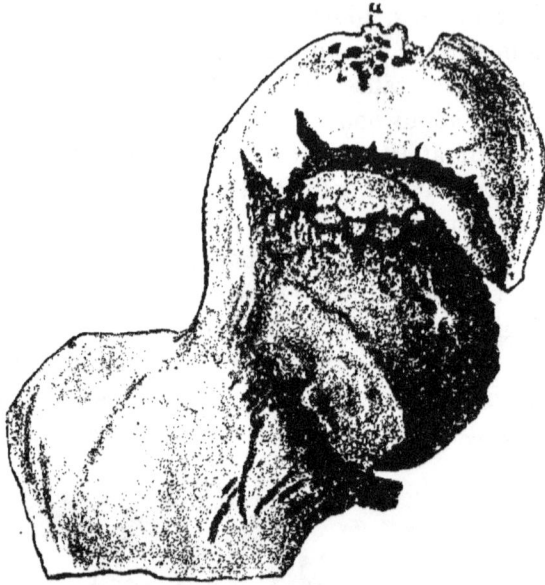

FIG. 81.—Cone-shaped sequestrum abutting against the cartilage, which is perforated (a) with numerous holes and tilted away from the whole head of the bone. (From Krause.)

recover so completely that we are still in doubt as to the location of the primary lesions. In the children of syphilitics, in cases directly traceable to injury, and in those cases that follow the infectious diseases there

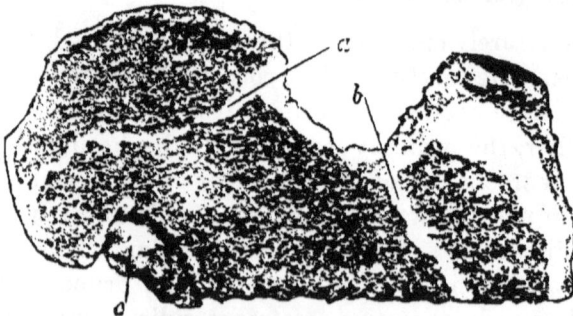

FIG. 82.—(a) Epiphyseal cartilage of the femoral head; (b) cartilage of the greater trochanter; (c) a cheesey sequestrum at the lower surface of the neck which has infected the joint. The head has been partially destroyed and is covered with a layer of granulation-tissue. (From Krause.)

is almost an equal uncertainty as to the primary patho-
genic process; but from the practical standpoint of
symptoms, prognosis, and treatment it is with a tuberculous arthritis that we have to deal.

Symptoms.—Almost invariably the first symptom
noticed is a slight limp. This usually begins before
there has been any complaint of pain, and it remains until a cure has
been effected, and not infrequently it persists to
the very end of life. In
a few cases there may be
one or more intervals of
intermission during the
early months of the dis-
ease. The limp is due to
the inability of the patient
to fully extend the thigh
on the pelvis more than to
any sensitiveness of the
joint to weight-bearing;
although in untreated
cases, and in very many
that are subjected to treatment, there comes a time
when walking is difficult
or impossible from the
sensitiveness of the joint
and its inability to sus-
tain the weight of the

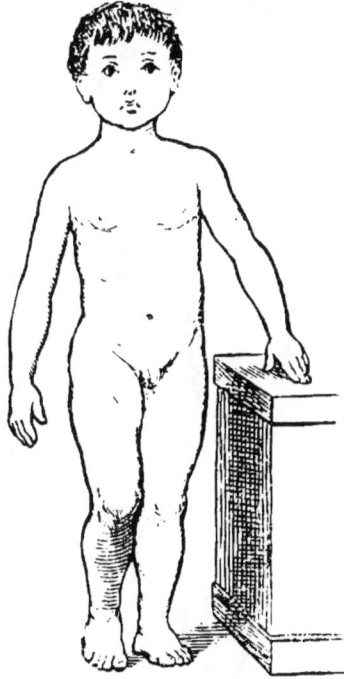

Fig. 83.—Early stage of hip-disease.
Flexion of right leg without noticeable lateral deformity.

body. Nevertheless, the characteristic limp of hip-
disease is not one of pain; it is rather one of impaired
function.

As a rule, the patient very early in the disease becomes restless in sleep, and may cry out without fully
awaking. The cry is peculiar, and consists in a sudden,

sharp, frightened scream, occurring during the first hours of sleep. These night-cries commonly precede and almost invariably accompany a period of pain. They have been considered as characteristic of osteitis, though there can be no doubt of the presence of osteitis in certain cases in which no night-cries are heard; it is more probable that they are indicative of a rather rapid development of a tuberculous abscess under tension. The symptom is valuable in the early stage of the disease only as corroborative of other symptoms and as a hint to the prognosis of abscess.

Pain is usually complained of at some time during the course of hip-disease, but the fact must not be lost sight of that it is rarely complained of until long after limping has appeared. It disappears months and often years before the joint is cured; and patients with nonsuppurative, and at times with suppurative hip-disease may never feel the least pain from the commencement of the limping all through the 3 or 4 years' course of the disease to a cure, resulting perhaps in a perfectly ankylosed joint. When pain is present it is usually complained of at the inner side of the anterior surface of the knee, but it may be felt in

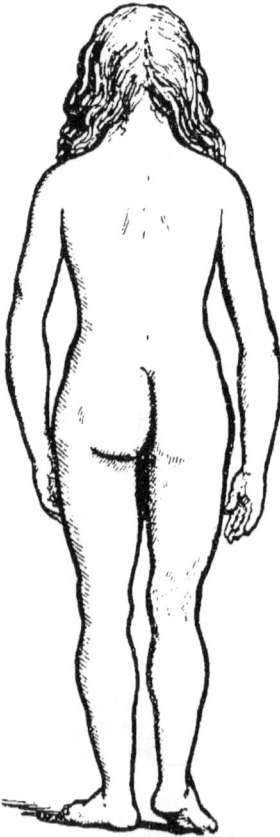

FIG. 84.—Flexion-deformity, without lateral deformity, showing obliteration of the gluteo-femoral crease.

any part of the hip or thigh ; like the night-cries it is
corroborative rather than a diagnostic symptom. It is
indicative of an osseous lesion, and, coming on or
growing worse without assignable cause during the
course of treatment, should be considered as pointing
to the development of abscess.

Inspection of a case of hip-
disease, stripped for examination, reveals flexion of the
thigh on the pelvis with or
without accompanying abduction or adduction. Most cases
present one or the other lateral
deformity, but the rule is not
invariable, and some cases run
their course and go on to a
great degree of flexion without
either abduction or adduction.

Abduction when present
usually is found in the early
stage of the disease, and has
been looked upon as indicative
of effusion into the joint, but
of this we do not feel certain.
We have observed abduction
late in the disease and in patients where we have never
been able to make out fluctuation. The position of abduction gives a false lengthening
to the limb, and is the cause
of the obliteration of the buttock-fold.

Fig. 85.—Early stage of hip-disease. Flexion and abduction
deformity showing false
lengthening.

Adduction usually appears when the flexion deformity has progressed so far as 25° to 40°. It gives a
false shortening to the limb and raises the buttock-fold,
making prominent the hip.

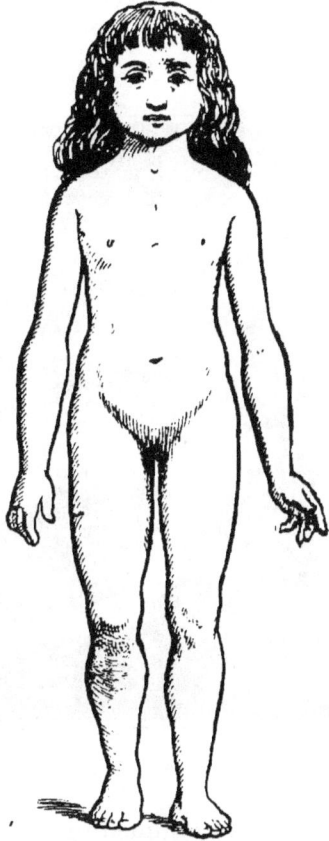

Outward rotation or eversion of the limb usually
accompanies abduction, and inward rotation accompanies adduction. Occasionally when there has been
great destruction of the upper end of the femur, outward
rotation will be found associated with adduction. The
cause of the malpositions at the hip have not been
absolutely determined ; but it would seem probable that
the position of the tuberculous focus, together with the

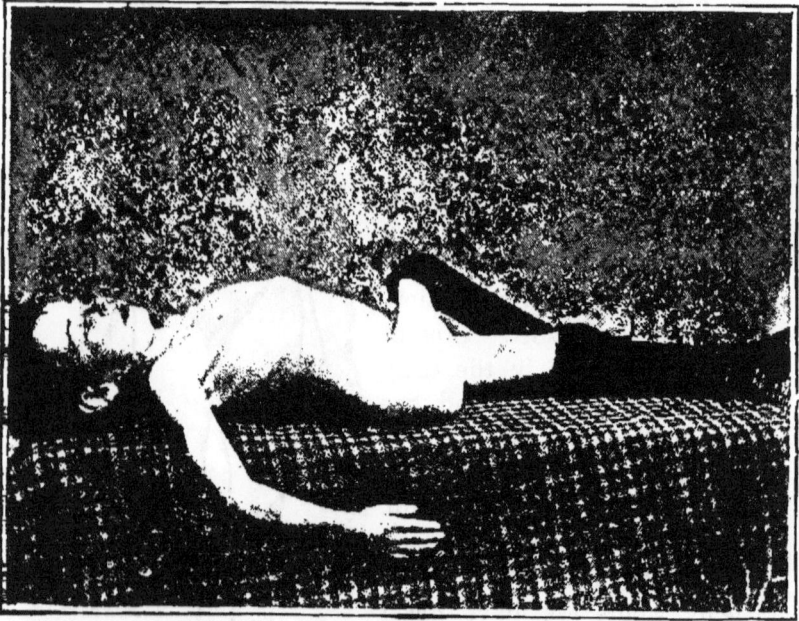

FIG. 86.—Disease of the right hip. Marked flexion and adduction deformity.

attitude assumed by the patient, determines the malposition. The patient assumes the attitude of greatest
comfort, whether lying, sitting, or standing, and this
strongly influences the deformity in the early stages of
the affection. Later the malposition is determined by
the unbalanced force existing between the opposing
groups of muscles in their effort to immobilize the joint,

or between the muscular groups on the one hand and the mechanical device employed in the treatment of the disease on the other hand.

Muscular shrinking, generally believed to be due to reflex influence, comes on early, is very constant, and may be regarded as a very valuable symptom. It has been claimed by careful observers that this muscular

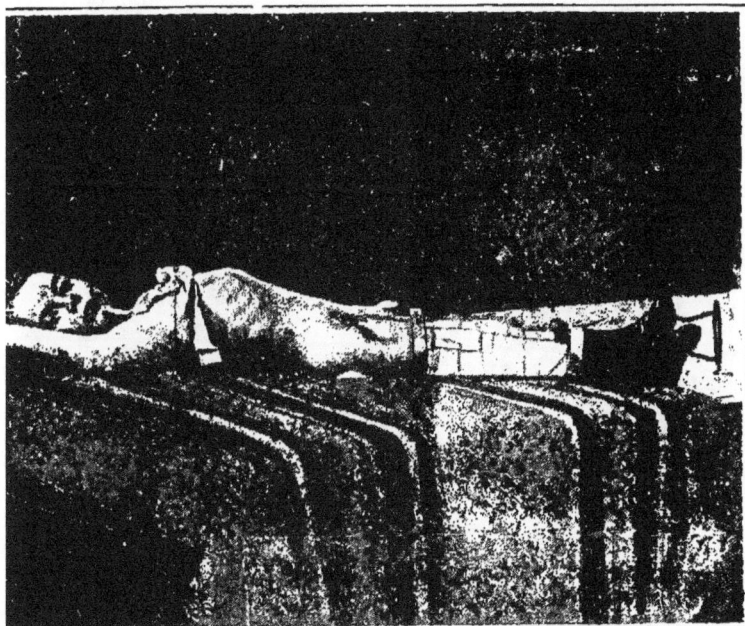

FIG. 87.—Same patient as shown in Fig. 86, with flexion and adduction-deformity reduced. Also shows scar from hip-abscess; also the Thomas hip-splint applied to the right leg and the high patten on the left shoe.

atrophy is due solely to disuse, inasmuch as it bears no constant relation to any other factor involved; but we have observed a patient who had limped for only 2 weeks, who had suffered no pain, and had not ceased from his usual avocations, whose affected thigh measured one inch less in circumference than the opposite thigh, a difference which went on increasing at a less,

but at an unusually rapid, rate during the succeeding
fortnight before the application of apparatus. The fact
that we do not know the relative sizes of the thighs
before the limping commenced, renders the case of no
positive value in controversial argument, but we are

Fig. 88.—Disease of left hip. Marked flexion, adduction, and outward rotation.

much inclined to believe that there is present in these
cases a shrinking of the muscular masses over and
above that due to disuse and to the constricting effects
of the dressings used. The muscular shrinking affects
the thigh first, later the buttock, still later the calf, and

in some instances the whole limb. The symptom is so
constant that it may be regarded as of very great diagnostic value. We have, however, seen one case in
which there was no muscular shrinking, but the thigh

FIG. 89.—Same patient as shown in Fig. 88, with deformity reduced. Shows
Thomas hip-splint with left wing of chest-band drawn higher than the right.

was actually larger than the other throughout the entire
course of the disease, which never presented any evidences of a suppurative process.

Shortening may be either actual or practical, true or

false. True shortening is due either to arrested growth, to actual bone-destruction, or to partial or complete displacement of the head of the femur from the acetabulum. None of these conditions are found early in the disease, when a false lengthening is often present; but an accurate record should be kept, for in this way the ultimate length of the limb can best be prognosticated. The true shortening is found by measuring from the anterior superior spine of the ilium to the inner malleolus on each side when the limbs are in like relations to the median line of the body. Practical, or false, shortening is due to adduction or flexion, or both. The false shortening is found by measuring from the umbilicus to the malleoli when the limbs lie side by side. It is from the relation of the actual to the practical shortening that the degree of adduction is calculated; and the relation of the true to the false lengthening enables us to find the degree of abduction. A limb may be really shortened, but by being abducted be apparently lengthened.

LOVETT'S TABLE.

Distance Between Anterior Superior Spines in Inches.

	3	3½	4	4½	5	5½	6	6½	7	7½	8	8½	9	9½	10	11	12	13
¼	5°	4°	4°	3°	3°	2°	2°	2°	2°	2°	2°	2°	2°	1°	1°	1°	1°	1°
½	10	8	7	6	5	5	4	4	4	4	4	4	4	3	3	3	3	2
¾	14	12	11	10	8	8	7	7	6	6	5	5	5	4	4	4	3	3
1	19	17	14	13	11	10	9	9	8	7	7	7	6	6	6	5	5	4
1¼	25	21	18	16	14	13	12	11	10	9	9	8	8	7	7	7	6	6
1½	30	25	22	19	17	15	14	13	12	12	11	10	10	9	9	8	7	7
1¾	36	30	26	23	20	18	17	15	14	13	13	12	11	10	10	9	8	8
2	42	35	30	26	23	21	19	18	16	15	14	14	13	12	12	10	10	9
2¼	...	40	34	30	26	24	21	20	19	17	16	15	14	14	13	12	11	10
2½	...	...	39	34	29	27	24	22	21	19	18	17	16	15	14	13	12	11
2¾	...	...	...	38	32	29	27	25	23	21	20	19	18	17	16	14	13	12
3	...	...	...	42	35	32	29	27	25	23	22	21	19	18	18	16	14	13
3¼	...	...	...	...	39	36	32	30	27	26	25	22	21	20	19	17	15	14
3½	...	...	...	...	...	40	35	33	30	28	26	24	23	22	21	19	17	16
3¾	...	...	...	...	...	...	38	35	32	30	28	26	25	23	22	20	18	17
4	...	...	...	...	...	...	42	38	35	32	30	28	26	25	23	21	19	18

(Left margin label: Difference in Inches between Real and Apparent Shortening.)

The difference between the real and apparent lengths of the limbs having been ascertained, we measure the

distances between the anterior superior spines of the
ilia, and then by Lovett's table compute the degree of
abduction or adduction. If the line which represents
the amount of difference in inches between the real and
apparent shortening is followed until it intersects the
line which represents the pelvic breadth, the angle of
deformity will be found in degrees, where they meet.
If the practical shortening is greater than the real

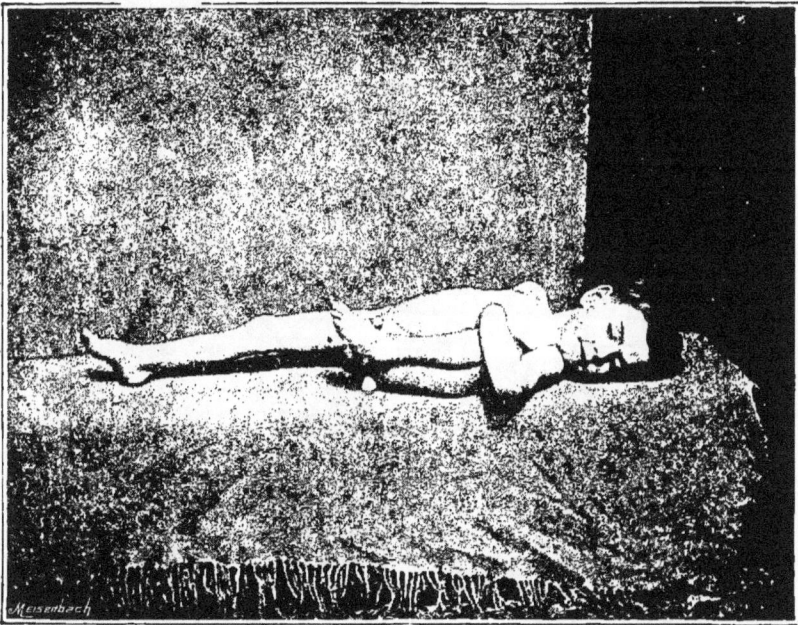

Fig. 90.—Same patient as shown in Figs. 88 and 89, cured with normal range
of motion.

shortening, the diseased leg is adducted; it less than
the real shortening, it is abducted. Take an example:
Length (from anterior superior spine) of right leg, 23;
left leg, $22\frac{1}{2}$; length (from umbilicus) of right leg, 25;
left leg, 23; real shortening, $\frac{1}{2}$ an inch; apparent shortening, 2 inches; difference between real and practical
shortening, $1\frac{1}{2}$ inches; pelvic measurement, 7 inches.

If we follow the line for 1½ inches until it intersects
the line for pelvic breadth of 7 inches, and we find 12°
to be the angle of deformity, as the practical shortening
is greater than the real, it is 12° of abduction of the
left leg.

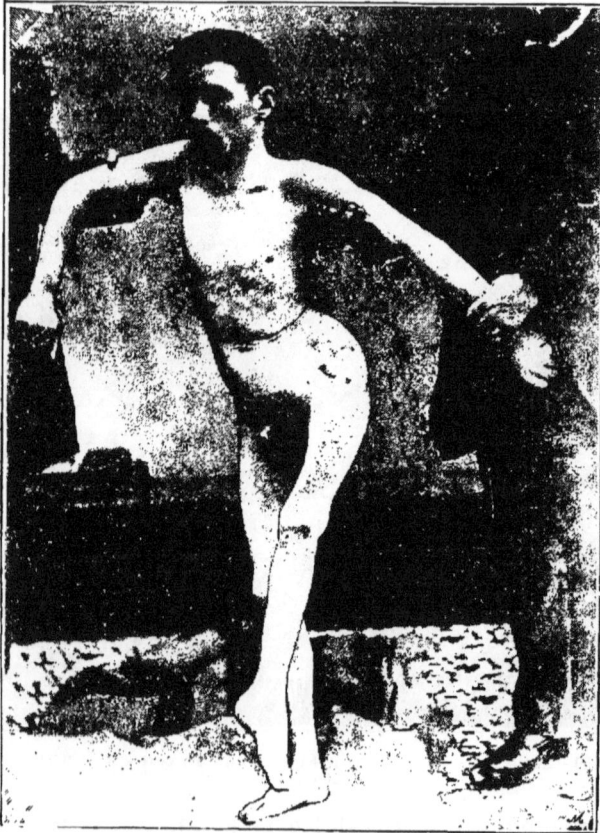

Fig. 91.—Disease of left hip. Untreated case. Flexion adduction, outward
rotation, and great shortening.

The angle of flexion may be estimated in the following manner: The patient lies on his back on a table;
the surgeon lifts the limb until the lordosis disappears
and the pelvis lies in normal relation to the trunk; he

then measures from the table along the thigh, following the line of the femur, for any distance, and from there drops a vertical line to the table, noting the length of both these lines. The decimal fraction obtained by dividing the length of the vertical line by the length of the line measured along the limb will give the sine of the angle formed by the oblique line and the table. By consulting a book of mathematical tables the angle is found. Kingsley measures a constant length of 24 inches along the thigh, and publishes a table showing the angle corresponding to the length of each vertical line from 1 to 24 inches.

KINGSLEY'S TABLE.

In.	Deg.	In.	Deg.	In.	Deg.	In.	Deg.
0.5	1	6.5	16	12.5	31	18.5	50
1.0	2	7.0	17	13.0	33	19.0	52
1.5	3	7.5	19	13.5	34	19.5	54
2.0	4	8.0	20	14.0	36	20.0	56
2.5	6	8.5	21	14.5	37	20.5	58
3.0	7	9.0	22	15.0	39	21.0	60
3.5	9	9.5	24	15.5	40	21.5	63
4.0	10	10.0	25	16.0	42	22.0	67
4.5	11	10.5	27	16.5	43	22.5	70
5.0	12	11.0	28	17.0	45	23.0	75
5.5	14	11.5	29	17.5	47	23.5	80
6.0	15	12.0	30	18.0	48	24.0	90

Actual lengthening of the diseased member rarely occurs, but we have observed it during the course of treatment by the long traction hip-splint. An accurate record of the angle of flexion and adduction, or abduction, is of the greatest importance. Upon a change in this angle, and upon this change alone at times, depends the diagnosis of disease; and upon it depends the diagnosis of a cure in all cases resulting in ankylosis. An ankylosed joint in which the angle of deformity is changing is not a cured joint; such a joint is capable, under proper treatment, of gaining a still greater degree of usefulness. In a case destined to result in ankylosis a cure is not effected until the angle of deformity ceases to change.

The involuntary muscular spasm which restricts the range of motion at the joint is the most important symptom of joint-disease. It is the first symptom to appear and the last to disappear, and it is the only symptom upon which dependence can always be placed in making the diagnosis. It is believed to be of reflex character, and to be due to irritation of the nerves that supply the joint. It affects only the muscles that con-

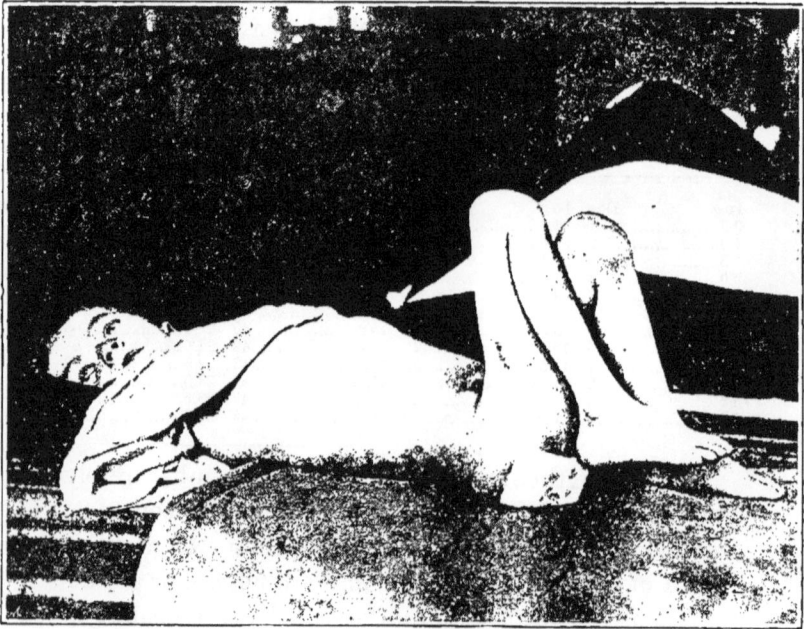

FIG. 92.—Flexion beyond a right angle; sinuses in the usual place.

trol the movements of the diseased joint, but it affects all of them. At times it is so slight that it can be recognized with certainty only by comparison with the healthy joint of the other side, and at other times it prevents all motion so completely that the joint appears to be ankylosed. Upon this symptom depends the diagnosis of the disease and the differential diagno-

sis from affections which closely simulate the disease.
Lack of normal extension and rotation is usually more
noticeable in the first weeks of the disease than are
restrictions to flexion and the lateral movements; but
a careful comparison between the movements possible
at the two hip-joints will render the defect apparent to
one who has become at all familiar with this peculiar
symptom. To describe the sensation which this invol-

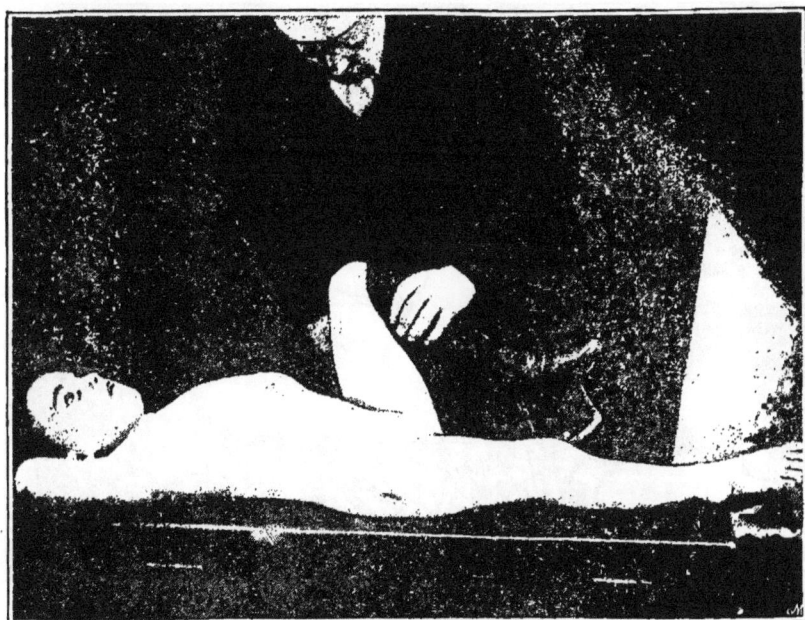

FIG. 93.—Scar from old sinus; flexion-deformity reduced.

untary muscular spasm imparts to the hand of the
examiner is scarcely possible; it is one of those things
better learned from the patient than from the teacher.
The restriction which involuntary muscular spasm gives
to attempted passive motion is not the slowly elastic
yielding of voluntary muscular opposition, nor the
sudden dead stop of ligamentous or fibrous adhesions;

nevertheless, it is a sudden and a positive stop, which, when once felt, will always be recognized.

The diagnosis of hip-disease is rendered comparatively easy by what is known as the Thomas flexion-test. This is founded upon our inability to extend an inflamed hip without producing lordosis. By lifting the knee of the sound limb until it touches the chest the pelvis is fixed and the spine is straightened. If

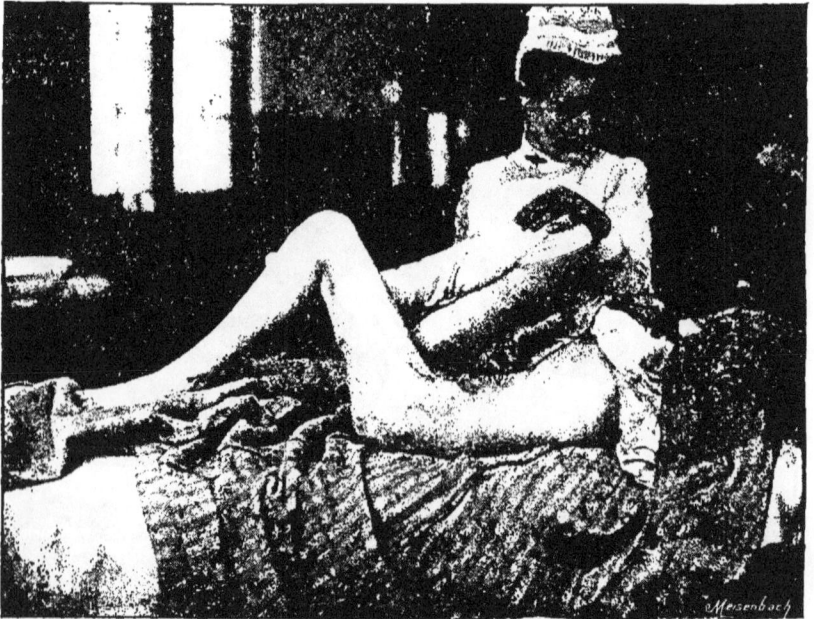

Fig. 94.—Shows flexion-deformity of diseased left hip when sound thigh is flexed on the body.

there be hip-disease the patient is unable to extend the thigh on the diseased side and it remains at an angle. If disease is absent the leg can quite easily be made straight. Few surgeons seem to have observed that if we take any healthy subject and lay him flat upon a table or other hard plane we can easily pass our hand under the lumbar vertebrae, but if we ask the subject to touch the table with his back he is able to obliterate the hollow without lifting his limbs. We have here, therefore, a very ready guide for the detection of deformity. In no case of hip-disease is the patient able to straighten his spine until art has stepped in and corrected the flexion-deformity.

The application of the flexion-test in the case of an infant requires considerable delicacy. A child 2 or 3

FIG. 95.—Same patient shown in Fig 94; flexion-deformity of left hip reduced.

years old is brought for examination. A vague history of irritability may be alone complained of, or pain may be occasioned when the child is washed. The surgeon is to find out in the first place whether there is an inflamed joint, and if so, on which side. The child is gently put upon the table, while the surgeon, without exciting alarm, holds a knee in either hand. The thighs are slowly flexed toward the chest, when it is observed that one easily yields to full flexion while the other becomes a little rigid. The stiff hip is then gently allowed to fall while the sound one is fully flexed. It will then be perceived that the diseased limb remains at an angle and cannot be fully extended. Stress must be laid upon the necessity of not startling the child and of not using the slightest force; while

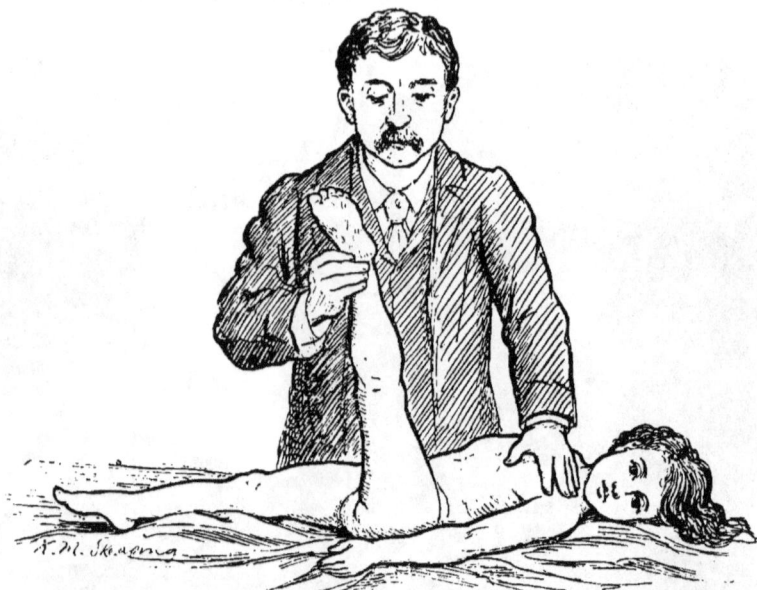

FIG. 96.—Testing the movements at the hip-joint.

care must be taken first not to flex the pelvis upon the spine, and secondly, to conduct the examination upon an even, flat surface. Although this test is not absolutely diagnostic, if the hip be complained of, and pelvic, vertebral, sacro-iliac and malignant disease be negatived, one can fairly infer the presence of coxitis.

The complications of hip-disease are abscess, spontaneous dislocation and separation of the head of the femur from the neck.

Abscesses occur in about half of all cases where treatment is not commenced very early. They may be present in any relation to the joint, but the most frequent position for the first abscess to appear is somewhat below and to the inner side of the anterior superior spine of the ilium. An abscess may appear early in the disease or at any time during its course. It is usually ushered in by a period of pain, night-cries, and increase of deformity; flexion is always present, and abduction is frequently found when abscesses appear early; adduction is more common when the abscess appears

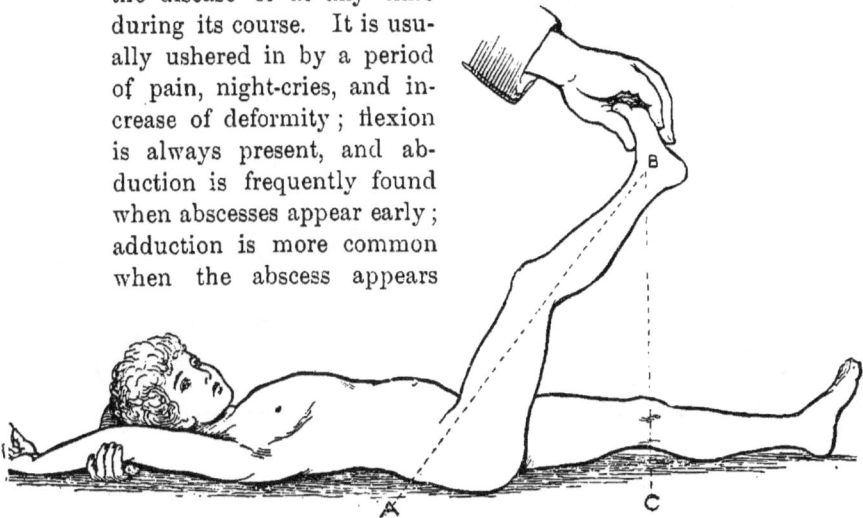

FIG. 97.—To measure the angle of flexion, the line A B should have followed the line of the femur instead of the lower extremity as a whole.

late in the disease. During the treatment of hip-disease any exacerbation of pain or tendency to deformity, unless there has been some well-recognized traumatism, is suggestive of the formation of an abscess and warrants that prognosis. The first objective sign is a brawny feeling in front of the joint; this is, or it soon becomes, tender to pressure, ultimately it softens in the center and fluctuation may be made out. The area of fluctuation extends, the extension being usually in the outward and downward direction, and at times fully two-thirds of the upper, outer and anterior portion of the thigh is occupied by the fluctuating tumor. The abscess, however, may appear posteriorly to the greater trochanter, or it may be first made out within the iliac fossa, where it has found its way through the acetabulum or through some of the natural openings of the pelvis; from here it usually makes its way up over the brim of the pelvis, following much the same course as a psoas abscess, and on reaching the thigh occupies the anterior and inner aspect. Sometimes an anal abscess

FIG. 98.

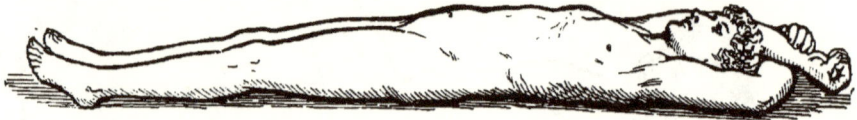

FIG. 99.

FIG. 98.—Shows the normal arching of the spine in a healthy person, which can be voluntarily effaced, as shown in Fig. 99.

is simulated, and after spontaneous opening the persistance of symptoms strongly suggests fistula in ano. Much care is required in differentiating it.

The course of these abscesses in untreated cases is towards spontaneous opening and evacuation. Rarely is there any especial fever or other constitutional symptoms except such as may be attributed to the pain; and pain is felt only while the abscess is intracapsular or subperiosteal; when once the pus escapes from the bone or joint the suffering ceases and almost invariably the general health of the patient improves.

In patients where the joint receives full protection these abscesses, even after they have attained a very considerable size, frequently disappear by gradual absorption.

Spontaneous dislocation, which occurs but rarely, may be due to distention of the joint or to destruction of the upper border of the acetabulum. These cases present the usual characteristics of a hip-dislocation, but the Röntgen picture will render the diagnosis certain.

Separation of the head of the femur from the neck is in our experience of such rare occurrence that we are unwilling to suggest any certain symptoms as characteristic.

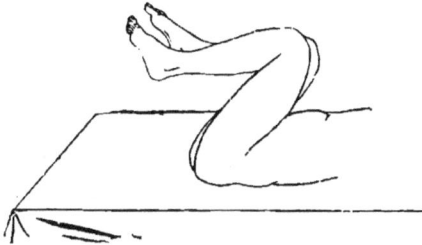

FIG. 100.—The diagnosis of hip-disease in an infant. Disease in left hip. Lack of perfect flexion.

The differential diagnosis of hip-disease is mainly from lumbar spondylitis, sacro-iliac disease, hysterical hip, congenital dislocation, traumatic dislocation, fracture of the femoral neck, and coxa vara.

In lumbar spondylitis the patient may walk with a limp because the thigh is held flexed, and motion is restricted in extension and in rotation during extension; motion is free in flexion and in rotation during flexion. In hip-disease the attitude may be the same, but action is restricted in all directions.

From sacro-iliac disease hip-disease is excluded by the same symptoms.

The hysterical hip may or may not present deformity, but rarely in a regular way and commensurate with the other symptoms; the rigidity is excessive, but more characteristic of voluntary effort than of involuntary muscular spasm; there are shifting superficial sensitive areas without evident cause; position changes during sleep; the limb can be carried through the full normal range of motion if a persistent effort be made and the attention directed elsewhere; and there is no muscular atrophy even when the limb has not been used for a long period.

Unilateral congenital dislocation at the hip gives a history of limping from the commencement of walking

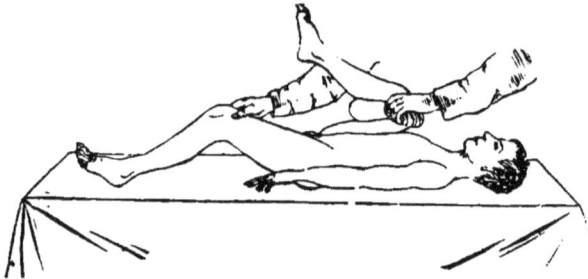

FIG. 101.—The Thomas flexion-test position; elbow hooked through the knee and forearm carried across the chest. The affected limb can not be forced down upon the table.

without pain or other disability; there is shortening of the limb of from ¾ to 2 inches when measured from the anterior superior spine of the ilium to the inner malleolus, but no shortening when measured from the greater trochanter to the outer malleolus, and the greater trochanter is found to be as far above Nelaton's line as the shortening shown by the first measurement. Motion is somewhat restricted in extension, abduction and outward rotation, but there is no characteristic involuntary muscular spasm. The head of the femur can usually be felt lying under the gluteal muscles on the dorsum of the ilium when the femur is rotated inward. The Röntgen picture makes the diagnosis conclusive. In hip-disease this amount of shortening would be associated with great, probably complete, rigidity to motion in all directions.

A traumatic dislocation at the hip would give the history of injury, with immediate and continuous dis-ability, growing better, if anything, rather than growing worse; the thigh would be held flexed, adducted, and rotated inward; there would be restriction to motion in all directions, and sensitiveness on motion without the characteristic involuntary muscular spasm of hip-disease. There is shortening from $\frac{3}{4}$ to 2 inches. The head of the femur can usually be made out under the

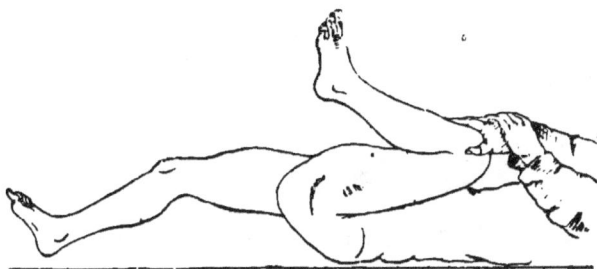

Fig. 102.—Tilting of the pelvis by extreme flexion, giving a false flexion to the right thigh in a healthy subject.

gluteal muscles, and the Röntgen picture renders the diagnosis positive.

Fracture of the neck of the femur, although a misfortune usually looked for only in those past middle age, does occasionally occur in children. There should be a history of injury with complete disability coming on either immediately if the fracture is complete, or within two or three weeks if the fracture is incomplete, the disability unchanging or slowly improving but never growing worse. There is shortening of $\frac{1}{2}$ an inch or more, and this is all above the greater trochanter as in dislocation. The limb is usually slightly flexed, somewhat adducted, and considerably rotated outward. The range of motion and the sensitiveness on motion depends upon the time since the injury and the amount of union that has taken place. There is no pain when the limb is at rest, and no tendency to an increase of the deformity. The Röntgen picture renders the diagnosis positive.

Coxa vara comes on either during the period of infantile or of adolescent rickets. There is a considerable period, often covering many months, of gradual shortening without other disability, unless there should be

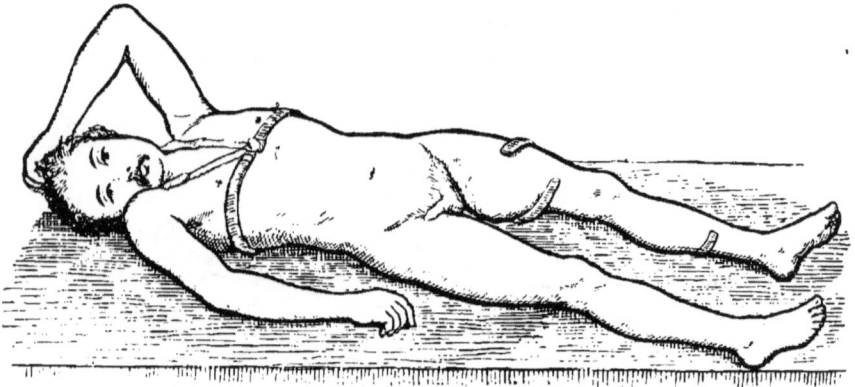

FIG. 103.—Abscess on the inner side of the thigh coming down from within the pelvis.

some traumatism, when there may be some sensitiveness and rigidity for a time. After the shortening has advanced to $\frac{3}{4}$ of an inch or more there is restriction to abduction, and later on, to extension, and finally to all motions at the joint. The shortening is all above the greater trochanter, and this can be demonstrated as in fracture of the neck and in dislocations. The characteristic involuntary muscular spasm of hip-disease is wanting. The Röntgen picture will clear up a doubtful diagnosis.

Treatment of Hip-Disease.—The principles governing the treatment of hip-disease are based upon a consideration of that joint both in health and disease. It
is the function of a normal hip-joint to permit of motion in several directions and to sustain the weight of
the body, both during walking and while standing at
rest, without injury to its structure. When a joint be-
becomes diseased, these functions become restricted or
abolished, motion is no longer possible, or possible to
only a limited degree, and the joint refuses to sustain
the superincumbent weight for any prolonged period.

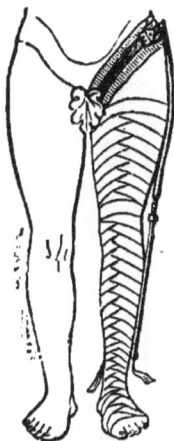

FIG. 104.—The Henry G. Davis hip-
splint. Designed to give elastic
traction ; to give protection during locomotion, and to allow " motion without friction " at the hip-
joint.

FIG. 105.—An early pattern of the
short Sayre hip-splint, with single
perineal band.

If we study the clinical evidences presenting at a
hip-joint as it passes from health to disease and 'back
to health again, we find them to be somewhat as follows: All the muscles whose function it is to move the
thigh on the pelvis gradually become more and more
rigid from involuntary muscular spasm until all motion

at the joint is abolished. The thigh becomes gradually
flexed on the pelvis, and usually at first abducted;
later on, as flexion increases, it becomes adducted; but
in either case the position is such, that in walking the
full weight is not thrown upon the diseased member

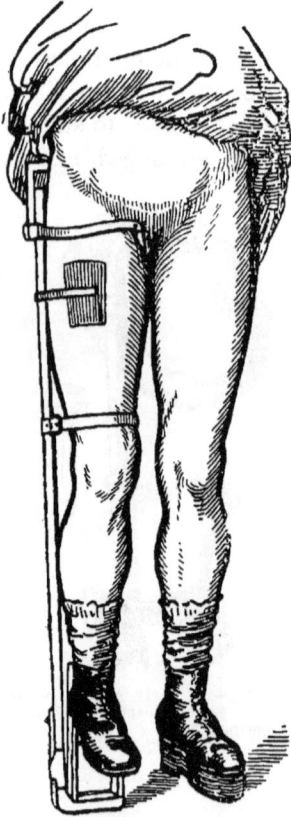

FIG. 106.—The Taylor splint.   FIG. 107.—The Judson long traction
hip-splint.

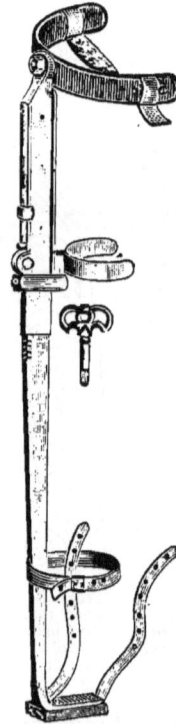

for more than a brief time at each step, and prolonged
weight-bearing, while standing at rest, is not possible.

The joint becomes more and more sensitive to the
vibration of locomotion, weight-bearing is no longer
tolerated, and the patient takes to his bed. The leg in

any case assumes the position of greatest comfort, and the muscular spasm protects the joint, which by this means becomes locked. As the patient falls asleep the muscular spasm relaxes somewhat, and if the limb does not lie securely fixed, motion takes place at the joint, injury is inflicted, the patient screams with pain, and the muscles are again on guard.

Long-continued malposition results in structural shortening of the tissues on the side of the flexion, and immobilization of the joint is then maintained with but little muscular effort. When the joint has been free from motion and weight-bearing for a certain time the tenderness passes off, and the patient is able to move about his bed without suffering, and ultimately arises and walks, often bearing his whole weight upon the affected member without pain. Nevertheless, muscular spasm and rigidity are maintained for a very considerable time. When the disease has terminated the spasm disappears, but the structural shortening of the soft parts remains and yields gradually to use during the subsequent months and years; but if the disease has been of a severe type it always remains to some extent.

The result of this cure by the natural process is usually a limb flexed and adducted with true or false shortening, and a joint which lacks the normal range of motion. These defects appear to be due to the prolonged course of the disease, which hinders the growth of the limb and renders more rigid the shortened muscles; to the position of deformity in which the leg rests while structural shortening takes place, giving rise to permanent flexion and adduction and to their result, false shortening, and finally to the exaggerated bone-erosion and consequent true shortening brought about by Nature's unaided imperfect immobilization and protection.

The efforts of Nature to effect a cure may be supple-

mented by art. The means which art adopts are: To
protect from deformity, or, if it has already appeared,
to correct it, and thus rob the muscular contracture of
its deforming power; to immobilize the joint, and thus
relieve the muscles from a state of spasm and subsequent contracture; to relieve pain and prevent the
bone-destruction due to both attrition and pressure; to

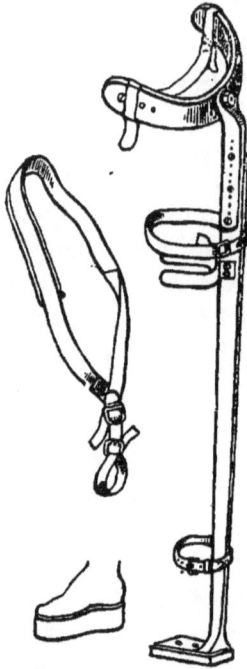

FIG. 108.—The Judson perineal crutch, with suspender strap and wooden patten at the side.

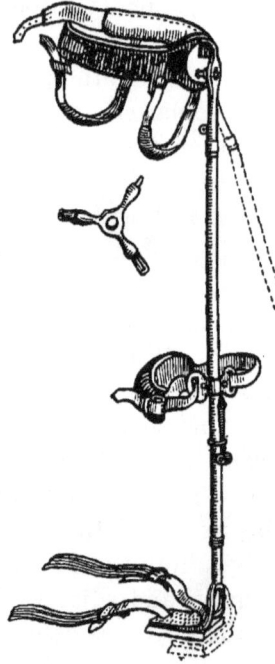

FIG. 109.—The long Sayre hip-splint showing action of abduction screw.

relieve the joint from weight-bearing and the pressure
arising therefrom ; and finally, to diminish all these by
shortening the course of the disease. The all-essential
element of treatment, beyond the correction of any existing deformity, may be summed up in one word—
rest. The ideal treatment would be perfect rest of the

joint from active and passive motion, from the jarring incident to all locomotion, and from intraarticular pressure due either to muscular spasm or to weight-bearing. Such an ideal treatment we do not think has ever been attained.

The fathers in surgery treated hip-disease by rest in bed and by more or less successful attempts at immobilization. Decros, in the *Gazette des Hôpitaux*, of June 30, 1835, published a case of hip-disease in which traction was employed. In 1839, J. H. James, of Exeter, Eng., presented at the Provincial Medical and Surgical Association, at Liverpool, a plan of immobilization for the treatment of fractures of the thigh by the use of traction in the axis of the shaft of the femur. In the same year William Harris, of Philadelphia, published a series of 4 cases of hip-disease in the *Medical Examiner*, January 19, treated by traction and countertraction, combined with Hagedorn's apparatus for fractured thigh. The first of these cases was treated 4 months after the publication of Decros' paper. The first portable traction hip-splint was devised by Ferdinand Martin, and is illustrated in Bonnet's *Therapeutics of Articular Disease*, Paris, 1853.

Following these, traction was used by various surgeons with weight and pully and other devices, as a means of immobilization in hip-disease. In 1859, Henry G. Davis, then a resident of New York City, presented a plan of treatment essentially different in principle from any that had been previously employed. It consisted of a mechanical device, intended to give elastic traction and counter-traction at the hip-joint, without restriction of the normal motions of that articulation, the attempt being to separate the articular surfaces and to thus obtain " motion without friction " at the joint. Another radical change in principle was that the apparatus was to be employed while the patient walked, it

being expected to furnish ample protection to the joint
from the traumatism of locomotion. These principles
of an ambulatory apparatus, which permitted motion
at the joint and protected it by elastic traction and
counter-traction, were at once adopted by Dr. Lewis A.
Sayre, Dr. Charles Fayette Taylor, and others, and the

FIG. 110.—The Shaffer hip-splint.   FIG. 111.—The Ridlon long traction
hip-splint.

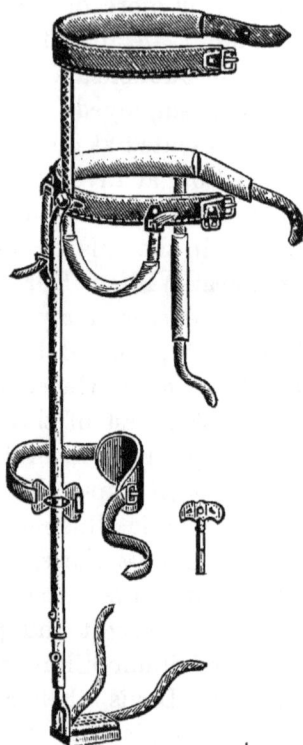

treatment, which was believed to allow " motion without friction " became known as the " American method "
of treatment.

It is, perhaps, unnecessary to say that the principles
upon which this treatment was based have been entirely
abandoned by the profession. Traction obtained by a

mechanical device, known as the long traction hip-splint, is still used both during recumbency and during locomotion, but it is no longer used with the idea that "motion without friction" is a mechanical possibility. Perhaps the best commentary upon the use of the long traction hip-splint is to be found in the fact that in the city of New York three of the veterans in the profession

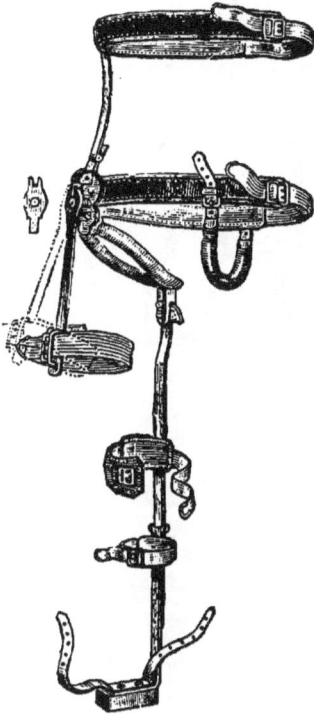

FIG. 112.—The Phelps splint.   FIG. 113.—The Phelps splint applied.

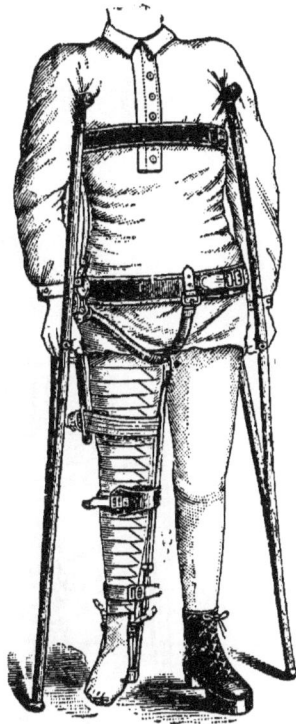

use practically the same splint in the same way for the accomplishment of three different ends, namely, Dr. Judson uses the splint for the fixation it gives; Dr. Sayre uses it for the protected motion which it permits, while Dr. Shaffer believes the beneficial effect chiefly rests in the traction which it exercises.

We would not be understood as denying that uninterrupted inelastic traction is an effective, though by no
means the most effective, agent for obtaining fixation
during recumbency; but traction applied by means of
an apparatus upon which the patient walks is quite

FIG. 114.—The Blanchard splint, for anteroposterior fixation, longitudinal
traction and lateral traction.

another matter. As was pointed out so long ago as 1879
by the late Dr. Joseph C. Hutchinson, of Brooklyn, it
increases the up-and-down motion, a motion that is
quite possible in a disorganized hip-joint, with each step

taken. It is only necessary to observe a child walking upon a long traction-splint to recognize this fact. The splint is applied while the patient is recumbent, and is made to exert a traction-force of from 8 to 10 pounds; the weight of the splint is from 4 to 8 pounds; when the patient stands upon the healthy extremity and lifts the affected member and the splint the traction upon the joint must be from 12 to 16 pounds; in taking the next step the splint is placed upon the ground and the sound limb lifted, bringing the whole weight of the patient to bear upon the perineal supports; they yield somewhat, the splint bends a little, and the traction-

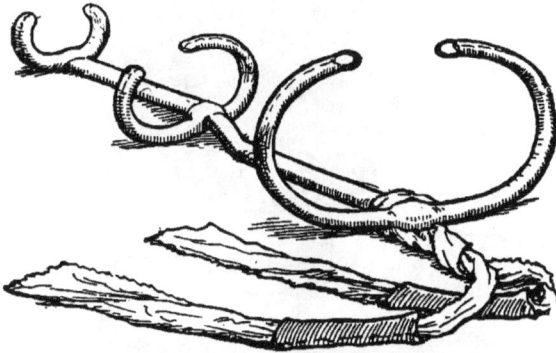

Fig. 115.—The Thomas hip-splint, fitted with shoulder straps.

force is entirely relaxed, as is shown by the straps bagging at the ankle; with the next step, when the splint is lifted from the ground, 15 pounds of traction is again in force. We have thus an alternate traction downward of 15 pounds and a relaxation giving full sway to the upward pull of muscles dominated by involuntary spasm. This push-and-pull or pumping action at the joint goes on with each step in walking, or at the rate of about 3,000 strokes an hour as the child runs about in his ordinary play. Looked at theoretically this method would appear to be a most effective means for destroying the joint, but as a matter of fact most of the cases treated in this way do marvelously well. To be sure, when treated by a traction-splint of the usual pattern (Sayre, Taylor, or Judson), that does not rise above the pelvis, and, as Lovett long since pointed out, allows about 30° of anteroposterior motion, most of the cases recover with flexion and adduction deformity, and with both true and false shortening, and much rigidity.

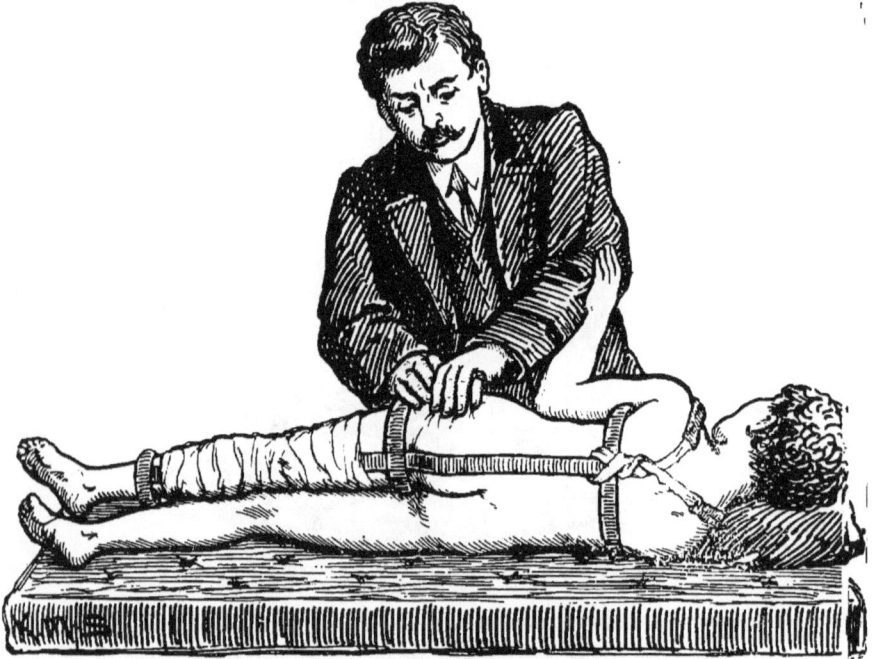

FIG. 116.—Method of changing the line of pressure on the skin from the Thomas hip-splint.

When the long traction-splint is supplemented by a body-piece, as in the Ridlon and the Phelps splints, with no motion between body-piece, hip band, and extension-bar, the splint becomes the most useful means for the correction and prevention of lateral deformity, and is second only to the Thomas hip-splint for correction and prevention of anteroposterior deformity. Trac-

tion, during recumbency, when combined with leverage
is a most effective means for reducing deformity, and in
certain sensitive cases during the development of
abscess, is a very efficient aid in allaying muscular
spasm, and reducing the paroxysms of pain. In other
equally sensitive cases it is not well borne, and positively
increases the suffering. In the majority of cases it is
neither indicated nor contraindicated. In cases that
are no longer particularly sensitive it may be used as a

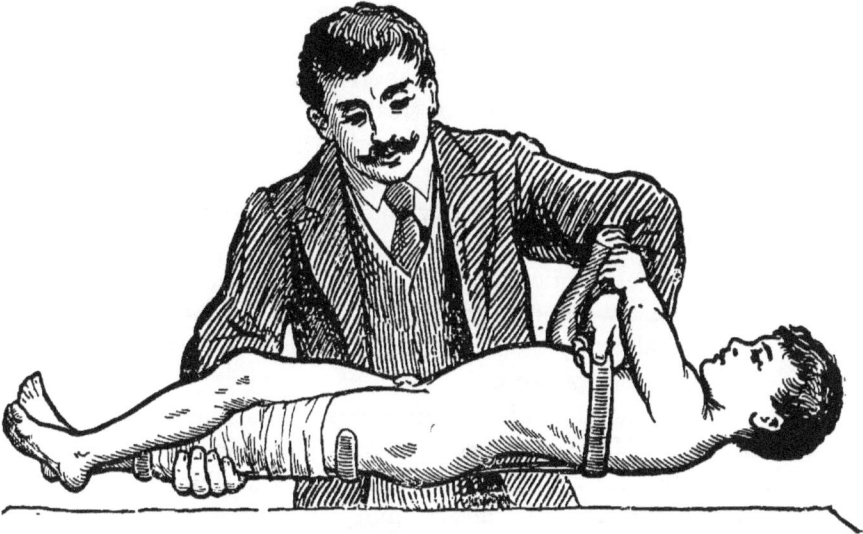

Fig. 117.—Method of lifting a patient in the Thomas hip-splint.

walking-brace without apparently doing harm, despite
the theoretic evidence to the contrary, in the vast
majority of cases. In patients too young to be trusted
with axillary crutches, it furnishes the safest protection
which we have against weight-bearing. The splint is
expensive, costing about four times as much as the
Thomas splint, and it requires intelligent care on the
part of the parents, and frequent attention on the part
of the surgeon.

The long traction hip-splint of the Sayre, Taylor, or
Judson pattern we do not use except during convalescence. It does not, at best, readily overcome deformity;
it permits the development of a marked flexion with
abduction, or adduction, in cases where no deformity
at first existed; it does not prevent exacerbations of
pain and the development of abscesses; in a word, it
does not immobilize the joint sufficiently to give the
protection required during the active stage of the disease. With the body-piece added, as in the Ridlon and
the Phelps forms of the traction-splint, combined with
rest in bed for a longer or shorter period, the splint
proves an efficient means of treatment.

As has already been said, traction is not essential to
the successful treatment of the majority of diseased
hips. All that is necessary may be obtained by the
use of an apparatus of simple construction, of slight
cost, easy of application, not readily misplaced, rarely
requiring attention, and more efficient in reducing
flexion deformity than the traction-splint. We refer to
the Thomas hip-splint.

Before the time of the late Hugh Owen Thomas, of
Liverpool, Mr. Hilton used a somewhat similarly shaped
splint for the reduction of deformity; and since that
time Blanchard, of Chicago, and others have used a
somewhat similarly shaped splint. Most surgeons, however, have lost sight of the essential principles of construction and of use of the Thomas splint. The essential principle of construction is that it be made of soft
iron of a thickness that cannot be bent by the patient,
but can readily be bent and fitted by the wrenches of
the surgeon. The essential principle in the treatment
is that the brace be so applied that it absolutely prevents anteroposterior motion at the joint. Splints of
this general pattern when made of steel cannot be accurately fitted because of their elasticity; if made heavy

they cannot be bent even with the aid of wrenches; if made light the vibration rendered possible by their springiness counteracts all the beneficial effects of immobilization. Splints that are curved to follow the outline of the patient, like the Blanchard splint, lose much of their immobilizing force from lack of the greatest possible leverage. Lateral traction, as illustrated in the Blanchard splint, and in the Phelps traction-splint, appears to us to be an absurdity. That the lateral traction exerted by these two splints is an added means of immobilization we will not deny; but we do deny that

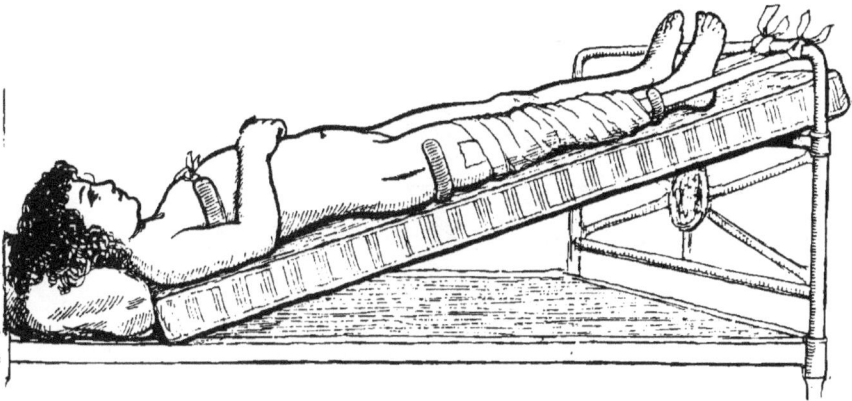

Fig. 118.—Fixed traction in bed used in combination with the Thomas hip-splint.

they act in any way to distract the head of the femur from the acetabulum; and it appears to us that this fact should be self-evident to any orthopedist or to any anatomist.

The Thomas hip-splint consists of a main stem, a chest band, a thigh band, and a calf band, and occasionally an abduction or adduction wing passing around the flank. The splint is constructed of the softest and toughest iron. Annealed steel is not the material to be used, inasmuch as sufficient rigidity cannot be obtained without rendering the parts too difficult to easily mould to the contour of the patient.

Most serious results accrue from making splints too light, and the following practical instructions may be useful: For an adult of about 6 feet the upright should measure $1\frac{1}{4}$ by $\frac{1}{4}$ inch; for an adult of about 5 feet 6 inches the upright should measure $1\frac{1}{8}$ by $\frac{1}{4}$ inch; in a child of 10 the upright should measure $\frac{3}{4}$ by $\frac{3}{16}$ inch; for a child of 5 the upright should measure $\frac{1}{2}$ by $\frac{1}{8}$ inch; for an infant of 2 the upright should measure $\frac{1}{2}$ by $\frac{3}{32}$ inch.

The wings should be the same width, and of such thickness that they may be readily bent by hand. In

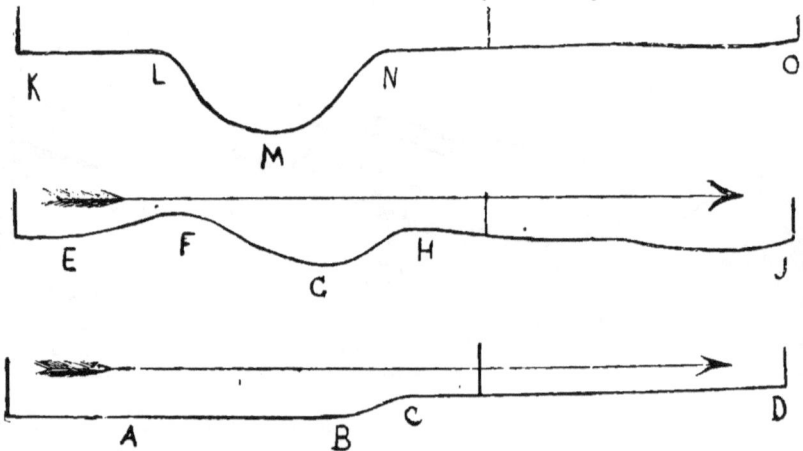

FIG. 119.—Correct and incorrect outlines of Thomas splints. A, B, C, D, correct outline; A, B, body portion; B, C, hip portion; C, D, leg portion. The two upper outlines are bent to follow the outlines of the patient to some extent; they lose all leverage and consequent effectiveness by the dip at G and M.

length the splint reaches from the lower angle of the scapula to the junction of the middle and lower third of the leg, passing down posteriorly to the hip-joint. In growing children it is customary to make that portion below the joint somewhat longer than that above, but nothing is gained in immobilization by making one part longer than the other, taking the hip-joint as the middle point. The upright stem is bent in two places, one opposite the fold of the buttock,

the other just above the joint, so that the leg-portion
and body-portion follow parallel lines distant from each
other from one-half to two inches, this distance depending upon the size and stoutness of the patient.

As a rule the stouter the patient the nearer do these
parallel lines approach each other. In case the trochanter is enormously hypertrophied the buttock bend
may be entirely dispensed with, and in that case the
body and thigh portions of the upright form a straight
line. The bends referred to should be rather rounded
than angular, as may be seen in the illustration. The
leg-portion from the fold of the buttock to the lower
end is perfectly straight, as is the portion from the

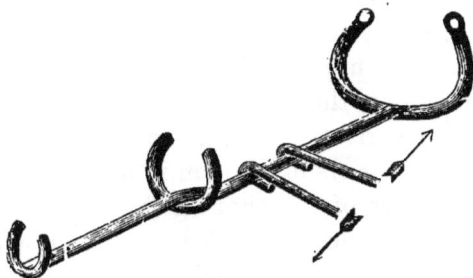

FIG. 120.—Showing method of twisting the main stem of the Thomas hip-splint.
This splint is being twisted to fit the left side.

bend opposite the joint to the upper end. The stem is
usually twisted somewhat in its longitudinal axis, so that
the body-portion lies slightly to the side and flat against
the curved outline of the chest, while the leg-portion
lies directly posterior to the middle line of the leg.
The buttock bend lies between the greater trochanter
and the ischial tuberosity. The twist varies according
to certain conditions which will be hereafter referred to
in describing the adjustment of the splint. The chest
band is also made of flat bar iron which varies in
width and thickness in proportion to the size of the
patient. It should be long enough to encircle the chest

to within an inch or two. If the splint be made reversible so that it may be applied to either leg, as is customary in hospital work, it is joined at its middle to the upper end of the stem, whereas, if it is to be used only for a certain individual case, it is joined from one to two inches to one side of its middle, so that one wing will be longer than the other. The longer wing encircles the chest opposite to the diseased side, its greater length being due to the greater distance it has to travel. The relative length of these wings may be determined by measuring from the lower angle of the scapula around each side to the front where it is intended that they should terminate. There is no special advantage in having these wings to end opposite each other, although it may make a somewhat neater-looking apparatus. The upper end of the main stem is forged flat and bent over the chest-band and the two are made fast by a single rivet. In each end of the chest-band a hole, three-quarters of an inch in diameter, is forged for the fastening of the shoulder-bandage, or, what is less convenient, holes are drilled for the attachment of a buckle and strap. The thigh band is made of flat iron, and, being placed on the surface of the main stem next to the patient, is joined to it by one rivet at a point about an inch below the lower bend. If the splint is to be made reversible the wings of this band should be made of equal length; if it is intended for one side only the inner wing should be made an inch or two longer than the outer. The calf band is also made of flat bar iron and is joined to the lower end of the stem by a single rivet in the same manner and with the same relative lengths of wings as the thigh band. When an adduction or abduction wing is required it is made from the same sized iron as the thigh band. This should be placed at such a point that it will pass around the flank midway between the crest of the

ilium and the ribs. This point is usually midway between the buttock bend and the chest band. The wings
are bent approximately to fit the imaginary patient,
and the surface of the entire splint next the patient is
lined with felt of one-fourth inch thickness. The whole
is then covered with that kind of sheepskin known to
the trade as basil leather or "tan sheep." This should
be put on wet and snugly stitched into place, so that as
it dries the shrinking will prevent any slipping upon
the iron. The stitching of this leather is, of course,
done on the surface of the splint away from the patient; it may be done with the so-called ball stitch, or,
what is more serviceable but less neat, the edges of the
leather may be drawn together and sewed through and
through after the manner of the harnessmaker with a
double waxed end; the redundant portion of the
leather is then trimmed off.

The splint is applied while the patient rests upon his
back, the wings upon the side away from the deformity
being opened out sufficiently to slide the splint under
the patient from the affected side without unnecessary
jar or movement. When the stem rests in place, the
leg-portion will be directly behind the middle line of
the thigh and leg, the part between the bends directly
at the back of the hip-joint, and the body-portion somewhat to the outer side of this line, the whole lying flat
against the chest, thigh, and leg. This fitting may be
done approximately with the hands, but better by
the aid of wrenches. The wings to the inner side of
the leg and thigh and the wing of the chest band on the
same side, namely, those on the side away from the
articulation, are drawn more closely than those on
the affected side. The reason for this is that the splint
tends somewhat to the affected side of the patient and
to draw the leg into abduction.

Particular attention must be paid to the bending of

the body-wings. As the chest is not circular it is
necessary that the body-wings should not be made circular, else intolerable soreness will result. By closely
examining Fig. 122, this arrangement will be seen. The
part between A and B upon which the patient lies is but
slightly curved and this allows the body to rest comfortably and travel easily towards the diseased side, B,
which is of great advantage. Another very simple expedient consists in making a small hole in the bandage
and passing it over the outer of the thigh wings and
rolling it under the splint and thigh and around the

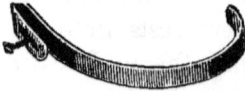

FIG. 121.—An adjustable abduction or
adduction wing for a Thomas hip-
splint.

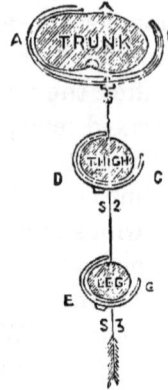

FIG. 122.—Showing the relations of
the bands of a Thomas hip-splint
to the chest, thigh, and leg.

limb, so that the appliance is pulled in the opposite
direction to that which it tends to travel.

When satisfactorily fitted a short piece of bandage is
wrapped around the splint and leg, pinned securely,
and another wrapped around the thigh above the knee,
or, what serves in some cases more satisfactorily, a
single piece is wrapped around the knee in the figure-
of-eight fashion and pinned with a large pin directly
through the covering at the back of the splint so that
the bandage cannot slip upon the splint, and any tendency of the splint to slip downwards is avoided.

If the splint is found too large, or, as when the patient grows, too small, it may be necessary to modify its length. This is quite simply done. If it is too large, draw the body-wings towards the abdomen; if too short, draw them towards the neck. A strip of broad bandage is then looped around the upper end of the stem below the chest band, and, having been twisted two or three times so that the ends will separate high up on the back, each end is carried over a shoulder and brought down to the hole in the end of the chest-band like a pair of braces; here each is tied securely, crossed to the hole of the opposite side and tied again, when the ends are firmly knotted. The final knot should be secured either with a long pin driven through it and twisted at its end, or with a bit of adhesive plaster.

FIG. 123.—Showing method of binding wings of chest-band upwards and downwards, and the method of looping a bandage over the outer wing of the thigh band to oppose rotation.

The splint should be applied without bending the main stem from the shape already described, if it is possible to force the leg at the knee reasonably near to the splint. The lumbar spine readily curves when there is flexion at the hip sufficiently to allow the limb to be brought down to the splint when there is as much deformity as fifty degrees; but if the deformity be very great, as much perhaps as ninety degrees, it may be necessary to bend the splint just enough to get the limb into contact with it when the fullest possible lordosis has been obtained. In these cases the bending is done at the upper bend of the main stem directly at the back of the joint. In practice, however, this will rarely be found necessary, and it has its disadvantages.

If there exists any considerable degree of abduction, a wing should be attached as already directed, passing around the flank on the side opposite to the disease. If there be any considerable adduction the wing is attached at the same point, but passed around the flank on the side where the disease is located. Care should be taken to draw these wings well in between the ilium and ribs, since pressure is not tolerated over

Fig. 124.—Showing extreme temporary lordosis produced in correcting extreme flexion deformity by anteroposterior leverage.

these bony points. At other times the body-wings are drawn toward the position taken by abduction or adduction wings when one cannot conveniently procure the additional wings. In the case of the very poor the hip-splint is often supplied by one of the authors (R. J.) without padding or leather. Lead-foil plaster is alone placed around the body-wings and stem. I the splint has been accurately fitted no sore or excoriation results.

If it is desired to prevent the patient from walking, a strip of iron is screwed on to the lower end of the splint, bent to pass free of the heel, and carried 10 or 12 inches below the foot, so that standing or walking is quite impossible. This piece is called a " nurse," and will be found, when children are restless, a safe precaution during the period of recumbency. If severe leverage be brought to bear over the buttock in order to reduce a marked and rigid deformity, care should be taken to shift the skin about twice a day where it presses with most force upon the stem, and to see that all parts about the hips are kept clean, dry, and well powdered, otherwise pressure-sores may result.

If the splint has been bent to fit the deformity, it must be straightened as soon as possible, sufficient opiate being given to quiet the pain during the few hours or days of the reduction of the deformity. During this time the patient must of course be kept in bed, and recumbency should be maintained until all pain and intense muscular spasm have subsided. When the deformity has been reduced the leg should be scarcely interfered with, the splint should not be removed, motion should not be tested, even the bandages at the knee should not be changed except they become slack or soiled. The most absolute quiet to the joint and to the patient must be enjoined, and the necessities of nature should be attended to by gently lifting both lower extremities and inserting the bed-pan. This can be done without causing pain in even the most sensitive joint. The good limb is placed gently across the diseased one, and the nurse lifts the patient by placing one hand under the splint just below the knee, while with the other she lifts the chest-band. When all pain, tenderness, and muscular spasm have been quies-

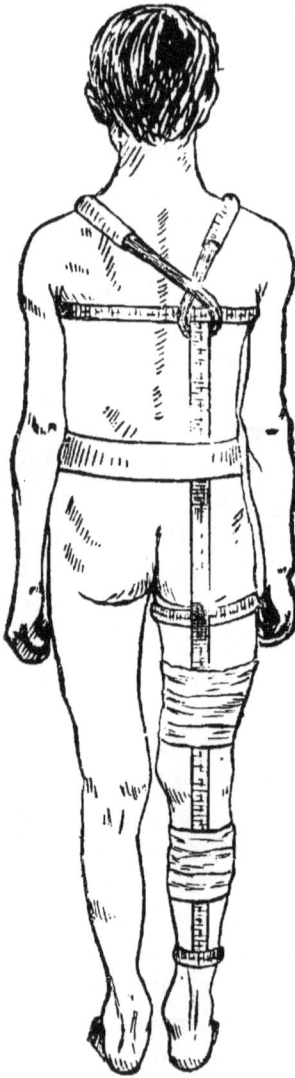

FIG. 125.—The Thomas splint, with
band around the hips, used as an
added means for immobilization
in certain sensitive cases.

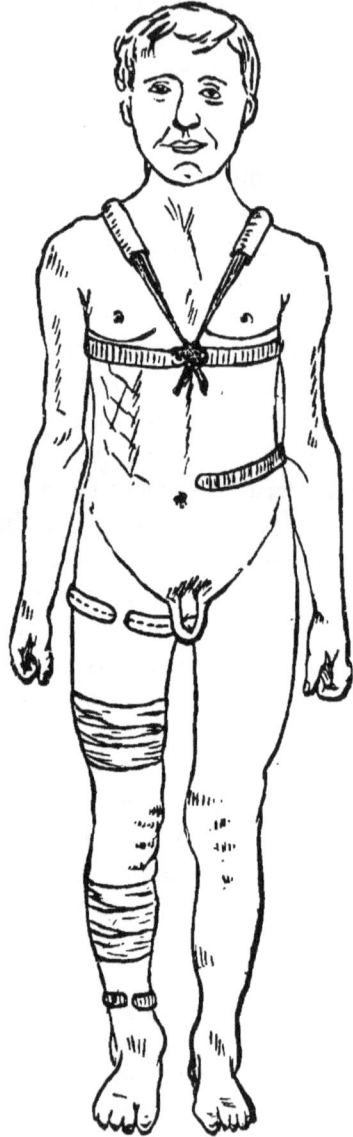

FIG. 126.—Front view of the Thomas
hip-splint with adduction wing.

cent for some weeks, and when no sign of fluctuation can be made out about the joint, the patient may be allowed to arise and get about on crutches, aided by a patten on the sound limb. The patten consists of an iron ring with two uprights, the ring resting on the ground and the uprights rising from the front and back, reaching to the shoe and fastened to the heel and sole. The ring is oval-shaped, and is made of square bar-iron not less than $\frac{3}{8}$-inch thick. It reaches from the ball of the foot to the middle of the heel, and its width is slightly more than that of the sole of the shoe. The uprights are of round bar-iron set at right angles to the plane of the oval ring, when viewed laterally, and slightly oblique when viewed anteroposteriorly, and should, at their lower ends, be welded to it; their upper ends are forged flat, pierced with three holes and bent forward. If the patten is to be attached to a thin-soled shoe it will be better to rivet these flattened ends to a metal plate shaped to fit the sole of the shoe and screwed to it. The height of the patten depends upon the size of the patient and should be from 4 to 6 inches, high enough to prevent the patient from reaching the ground with the toe of the affected side. This with the ordinary crutches completes the ordinary walking outfit.

For the most perfect result the patient should be kept recumbent until all pain, tenderness and muscular spasm have subsided. He may then walk about on crutches and patten until all the soft tissues about the joint are well atrophied, and all trace of the disease has disappeared. The patten may then be dispensed with and the crutches shortened, and in this manner he may go about for two or three months. If there be no evident return of the disease the crutches may now be thrown aside and the joint further tested by two or three months' use. All still going well, the

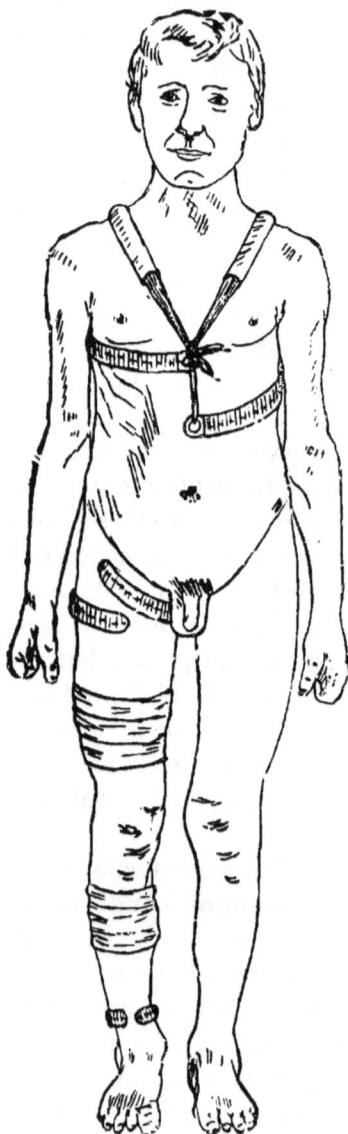

FIG. 127.—The Thomas hip-splint with the left chest wing drawn down to act as an adduction wing.

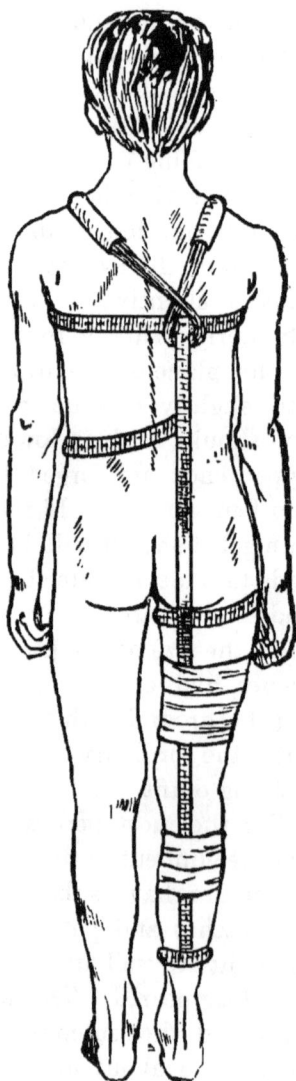

FIG. 128.—The Thomas hip-splint with adduction wing.

splint is cut off at the knee so as to permit flexion
there, a band being attached at the lower end after the
same manner as the calf-band. This short walking-
splint having been worn for two or three months, and
there being no return of the symptoms of the disease,
the splint is removed at night for a month or two. If
the joint remains well the splint is removed for certain
hours during the day, and then altogether, and the
joint finally tested for perfect cure.

The joint should be imprisoned long after the appearance of the disease has gone and after all subjective
symptoms have disappeared, for the sensations experienced by a patient recovering from articular disease
cannot be very reliable under the masking influence of
a splint. The test comes on removal of restraint, and a

FIG. 129.—The iron " nurse " that may be screwed to the
bottom of a Thomas hip-splint.

very critical time it is unless the surgeon has grasped
the knowledge whereby such a test becomes reliable.
No surgical textbook gives any allusion beyond vague
generalities to the means of knowing the right moment
to discard the treatment. There is no more danger of
relapse in cured joint-disease than there is of disease in
a healthy articulation. But if a joint be pronounced
fit for use when the remnants of inflammation have not
gone, it is easy to understand the very frequent refer-
ences to relapse which meet us everywhere. The law
may be again laid down: A joint is cured of disease
when the range of motion does not diminish by use, or
in those cases resulting in ankylosis a cure may be
pronounced where the angle does not change after use.

Plaster-of-Paris will be found a convenient expedient
in the treatment of hip-disease, and for some general
practitioners perhaps the most satisfactory method.

throughout the entire course of the disease. When used it should be applied from the ankle to the axilla, and it should be made especially strong opposite the hip-joint. A seamless, skin-fitting, combination garment is the best lining for the plaster; in the absence of this, the patient should be wrapped in bandages made from sheet-wadding. Bony points like the iliac spines should receive extra thick protection, lest pressure-sores develop. Any existing deformity at the hip should be corrected in so far as it is possible; and as the main existing deformity is sure to be flexion it is often difficult to decide how best to place the patient to minimize this deformity while the patient is being wrapped in the plaster bandages. Bartow, of Buffalo, is accustomed to partially suspend the patient, as in the application of the plaster jacket by the usual method, to stand the foot of the sound limb on a block so that the affected limb may dangle, this being steadied by an assistant who grasps the foot and makes slight downward traction. This we have found to be a very unsteady and trying position for young children, and to tend to swing the leg too far into abduction in all cases except those where adduction deformity is a positive feature. We are accustomed to rest the patient's pelvis upon a small support, raise his shoulders upon a pillow, and while traction is made from the foot by one assistant and from the shoulders by another, to apply the splint. In the few cases where traction upon the limb adds to the relief of the patient, we first put on a plaster stocking, and when it has set, continue the splint on upwards from this, taking care to carry it far up against the perineum, which has been previously protected by a broad strip of felt. In this way the traction exerted by the assistant is maintained to a great extent. The part of the splint opposite the joint may be strengthened by ribbons of wood or metal, by strips of wire netting, or by carrying the layers of

the bandage directly up and down at the front, the side
and the back, until the splint is an extra thickness at
this weak place. A plaster splint should be left on so
long as it is comfortable, unbroken, and fairly clean, for
it cannot be changed without disturbing the diseased
joint and usually doing it some harm. We rarely let a

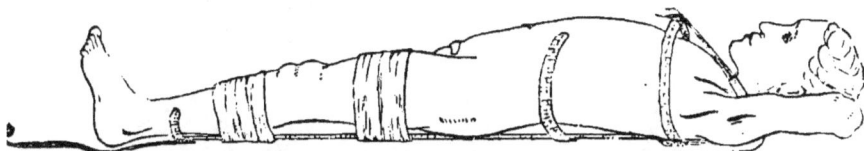

FIG. 130.—The Thomas hip-splint with abduction wing and "nurse" attached.

patient up while wearing a plaster splint. If he is
allowed up he should use crutches and a high patten as
in the Thomas splint.

All patients should be examined at the commencement of treatment for the purpose of diagnosis and record, and again at the close of treatment for a comparative
record and for the diagnosis of a cure. In each instance

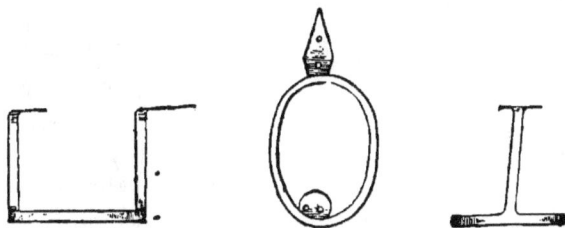

FIG. 131.—Three views of the iron patten, used on the sound side, in connection
with the Thomas hip-splint.

the patient should, if it is practicable, be entirely
stripped of all clothing, the attitude should be noted in
standing, walking, and lying, the amount of motion at
the articulation compared with that of the sound side,
the real and the false shortening, the abduction or adduction, the flexion and the atrophy.

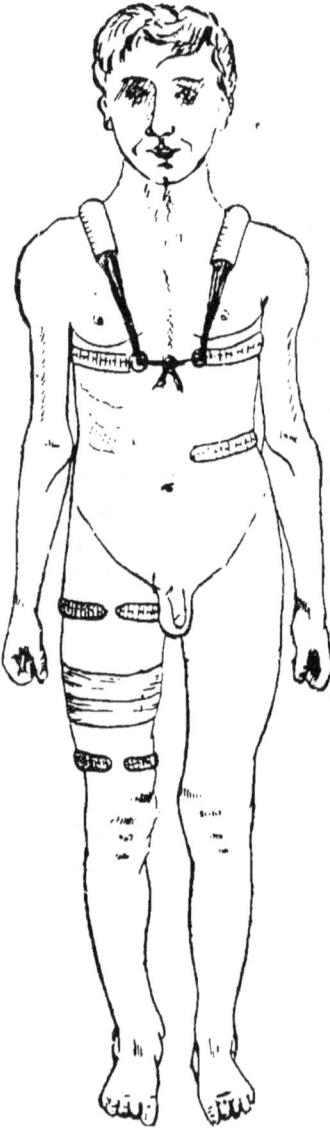

FIG. 132.—Front view of the short Thomas hip-splint with adduction wing. Used during convalescence.

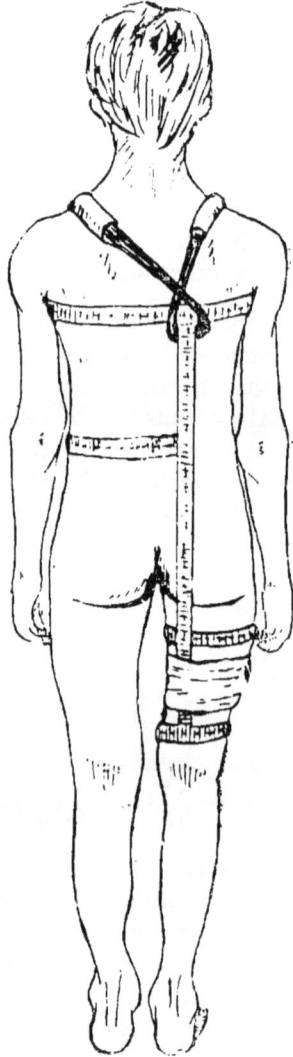

FIG. 133.—The short Thomas hip-splint with adduction wing. Used in convalescent cases.

It is our effort to encourage rather than otherwise the production of abduction in the cure of hip-disease. It diminishes the amount of practical shortening caused by displacement, erosion, or arrest of growth. By recognizing this, we are sometimes able, where there is perhaps two inches actual shortening, to slant the pelvis sufficiently to render the apparent or practical measurements equal on either side. Adduction, although often inevitable, should be energetically combated. Mr. Thomas used to sling a bag of shot around the pelvis, the weight being attached to the side it was desired to depress.

To render the pelvis flat in a definite position during examination it is customary for us to put the patient in what is known as Thomas' "flexion-test" position. This consists in flexion of the thigh of the sound side upon the trunk, so far that the elbow of that side can be hooked into the flexure at the knee and the forearm carried across the body. This gives a sufficiently definite position to render measurement made at different times by the same or different surgeons comparatively accurate, although unless the surgeon be careful, flexion to the Thomas position in some cases tilts the pelvis upwards and renders the record of deformity not only that of the existing flexion but also that of the amount of the normal extension.

These deformities may perhaps be more accurately measured with the goniometer, but the measurement with the tape is generally more convenient and is sufficiently accurate. The amount of muscular atrophy of both thigh and calf should also be recorded. This may be done by measuring the circumference of the limb at points similarly placed on both limbs. The degree to which motion is possible in the anteroposterior direction may be ascertained in the same way as the deformities are measured. The deformity having been

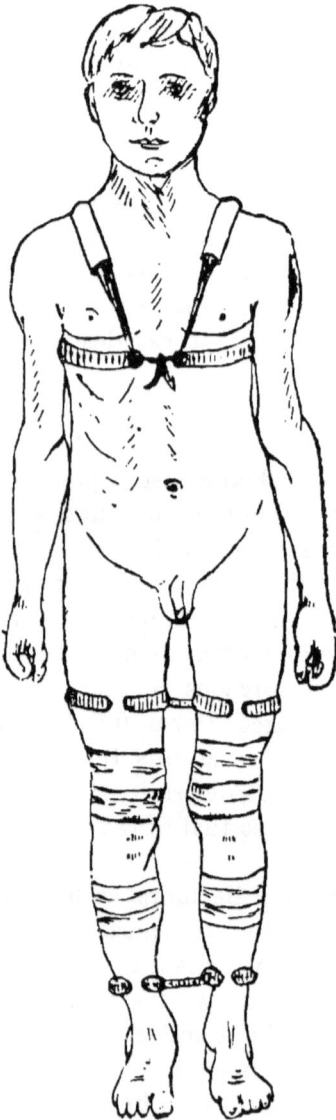

FIG. 134.—The double Thomas hip-splint, used in bilateral hip-disease, and in certain very sensitive cases of unilateral disease.

FIG. 135.—The double Thomas hip-splint, with girdle added about the hips.

measured and recorded, the presence or absence of
involuntary muscular spasm, limiting the motion at
the joint, should be tested. To make this test it is convenient to first test the leg of the sound side. The
pelvis is steadied by one hand, placed over the spines
of the ilium, while the other grasps the leg just below
the flexed knee; flexion, abduction, adduction, and
rotation are then tested. The affected limb is then
tested in the same manner. In timid patients or very
sensitive joints it may be as well to test rotation by
rolling the leg from side to side as the patient lies upon
the table, or the patient may sit with his legs dangling
over the side of the table and the foot may be swung
from side to side. The patient then is placed prone,
and, if the degree of deformity will admit it, the leg is
flexed on the thigh at a right angle, the ankle is grasped
by one hand, the pelvis steadied by the other hand
resting on the sacrum, rotation is tested by moving the
foot from side to side, and extension is tested by lifting
the whole limb from the table. The parts about the
joint should be palpated for tenderness, induration and
fluctuation. The presence and size of abscesses should
be noted, the location of sinuses and their character,
and the nature of the discharge. The advent of an
abscess is usually first ushered in by increased pain,
muscular spasm and increase of deformity. Tenderness may often be made out on palpation, and sooner
or later, induration or a boggy feeling is made mani-
fest. This usually is first to be felt directly in front of
the joint, although it may appear posterior to the greater
trochanter, or in fact at any point in the neighborhood.
As the abscess increases in size it usually extends downward and may come to spontaneous opening within
a few weeks, or not until after many months.

It does not appear to us that anything is to be gained,
while often much may be lost, by early operative meas-

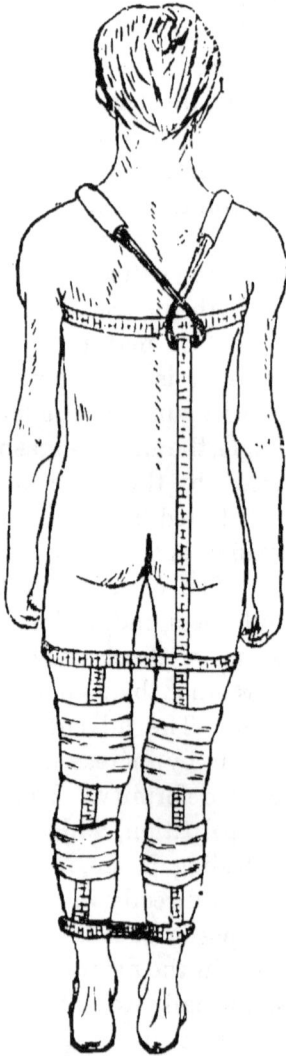

FIG. 136.—The double Thomas hip-splint modified to relieve the left hip from pressure.

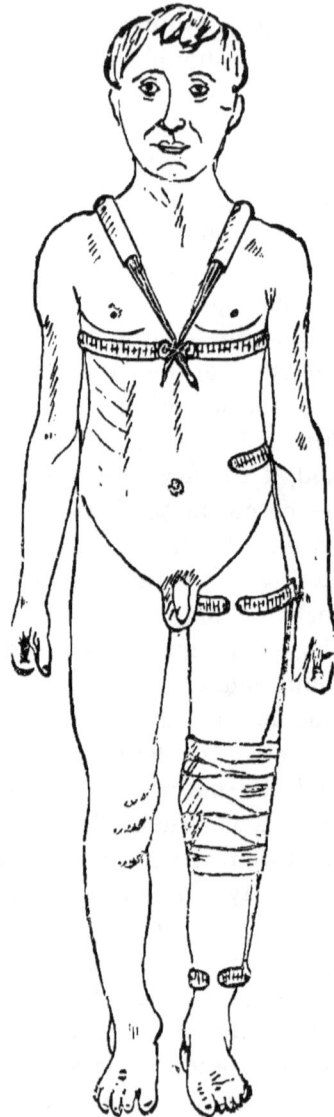

FIG. 137.—The Thomas hip-splint, with abduction wing and side bar to restrain tendency to in-knee.

ures for hip-disease, provided there are no constitutional
symptoms of septic infection. An abscess opened early
invaritably conducts to carious bone, and generally to a
joint extensively diseased. Rarely can all the tuberculous material be removed without a complete excision of
the upper portion of the femur and of the acetabulum.
Unless all diseased tissues are removed, a sinus is likely

Fig. 138.—The combination Ridlon
fixation-splint. This splint was
first used in a case of hip-disease
having a strong tendency to rotation of the limb. It may be used
for hip-disease and knee-disease
on the same side.

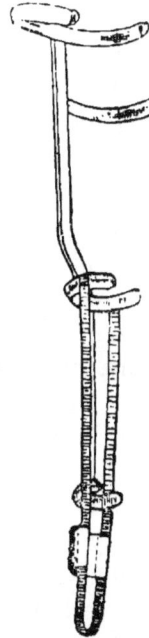

Fig. 139.—Loop attached to a Thomas
hip-splint for the treatment of hip-
disease and knee-disease on the
same side.

to remain which may subject the patient to septic infection. An abscess left unopened for some months
often descends a considerable distance and becomes cut
off from the original focus. In such a case careful
operative measures without drainage should result in

immediate closure and primary union, but it is not easy
to tell when the abscess is no longer connected with a
diseased bone or joint, and the surgeon who interferes
takes a very serious responsibility. He should not, in
our opinion, open such an abscess unless he can be
reasonably sure of removing all tuberculous material,
and of closing the wound without drainage. The use
of a drainage-tube leads to the formation of a tuberculous sinus, which is exceedingly difficult to heal, far
more difficult than a sinus resulting from spontaneous
opening. When operative measures are undertaken we
believe that the tuberculous tissues, whether sac of abscess,
wall of sinus, synovial membrane, cartilage, or bone,
should be removed by cutting with a knife or chisel,
instead of the scratching and scraping to which such
tissues are usually subjected by the so-called sharp spoon.
There can be no question that the risk of general infection is greater from a cutting operation than where
none is done, but the risk is much increased by the scraping process. In considering the treatment of these
abscesses it should be remembered that a very considerable number of them, if left to themselves, the joint
being put at complete rest, never go on to an opening,
but gradually dry up and disappear without any apparent ill effects to the health of the patient.

If it was certain that every abscess would come ultimately to the surface, or if there were any reason for
believing that the health of the patient suffered from
allowing them to remain unopened, or from their being
reabsorbed, operative measures would be justified in all
cases; but as there is no way of knowing what is to be
the course of any given abscess, we believe that the
indications for operative interference should be made
to depend solely upon the general health of the patient
and that no abscess should be opened unless the patient's
health be unquestionably suffering from the presence of

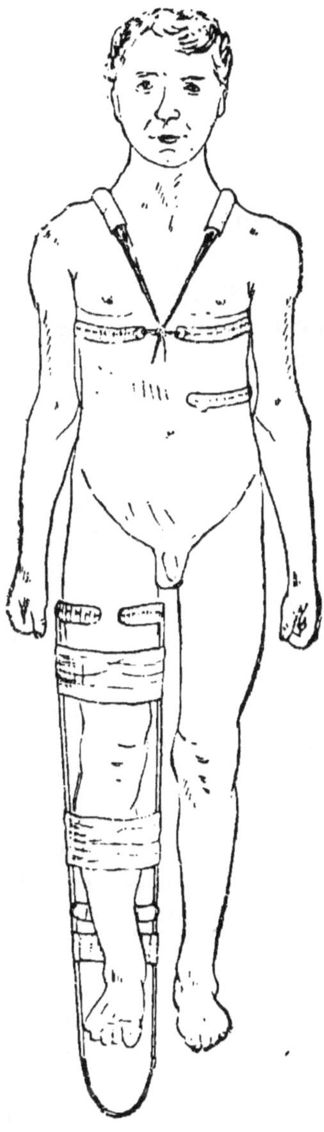

the diseased tissue. We would make the same rule regarding other operative procedures, such as the removal of the focus of disease within the bone, excision of the joint, and amputation. If any case in the recumbent posture should grow progressively worse under efficient immobilization of the joint, then the best obtainable hygienic operation for the removal of the disease would be indicated; but we have not seen such

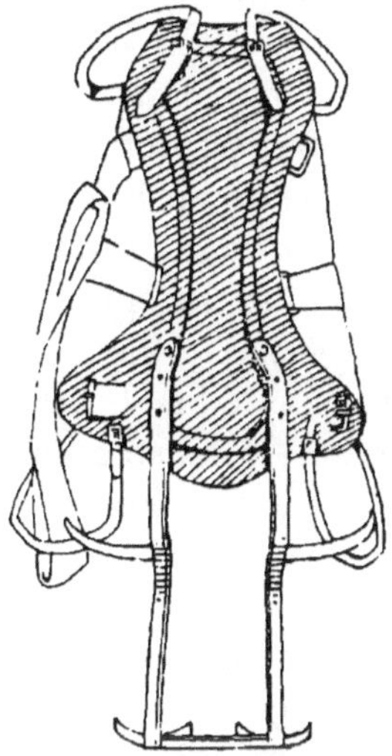

FIG. 140.—Front view of the Thomas hip-splint with adducting wing, and loop for the treatment of hip-disease and knee-disease on the same side.

FIG. 141.—Thomas' cuirass and double hip-splint combined; for the treatment of spondylitis and double hip-disease in the same patient.

a case except where the disease had been allowed to go on to an exceedingly advanced stage without any treatment whatever.

The preservation of the patient's life, then, we would make the only indication for an excision of the hip-joint, or for an amputation. The operation for excision, or for amputation, need not be described here, since they are found in all works on general surgery, but if excision be performed, thorough mechanical treatment following the operation is indicated and should be the same as in the treatment of any unsound articulation. This after-treatment we think is often neglected by the general surgeon, and may account for some of the relapses which have been reported. The mechanical treatment of these cases, consisting of immobilization and protection to the joint, should be continued until every evidence of unsoundness has been absent for a very considerable time. The hip splint, used after an excision, should be supplemented by the addition of fixative traction, as is done in some cases of fracture of the upper portion of the femur. A strip of adhesive plaster is applied to each side of the limb from the upper portion of the thigh to the neighborhood of the calf-band of the splint, the lower ends of these strips are then carried to the foot of the bed, or around the wings of the ankle-band, so as to secure the necessary traction, and fastened securely, and the splint is adjusted without the usual shoulder straps. While the patient is lying in bed without shoulder-straps the splint tends to work downward sufficiently to overcome the muscular contracture which would produce unnecessary shortening.

At times old and neglected cases will be presented for treatment with a serious deformity, and the question arises as to whether any operative measures are demanded. If muscular spasm be evident on attempting

motion at the joint even if there be no possible motion, or
if an apparently sound joint possesses a certain degree
of motion, the deformity can be corrected in a comparatively short time by the leverage action of the
splint; or the deformity may be corrected at once, or
nearly so, by anesthetizing the patient and placing the
limb in the best possible position. In these cases we
do not recommend section of the tendons, fascia, or
other contracted tissues, although there may be no very
serious risk in their division. To this subject, however,
we will return. We think it safer to divide the femur
with the chisel either through the neck or in the neigh-

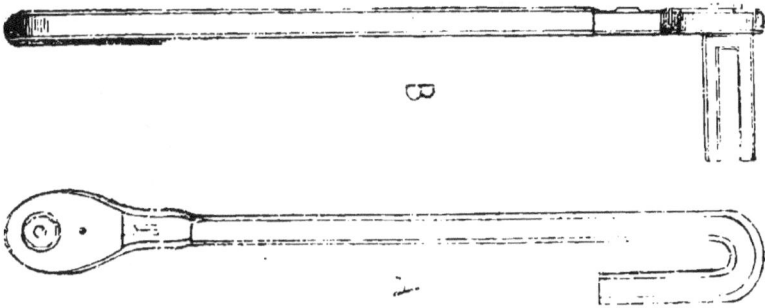

FIG. 142.—Wrenches for binding and twisting the Thomas splint.

borhood of the lesser trochanter than to attempt a
fracture by manipulation. The after-treatment of either
of these operations is the same as indicated after an
excision of the joint. The patient should remain in
bed until union is sound, when the splint may be removed and he may remain in bed, an equal time, without immobilization, or in place of recumbency in certain cases the splint may be cut off at the knee and the
patient allowed to go about with splint and crutches
without the patten, for a period equal to that which
was required for the union of the bone.

Cases occasionally appear with abscesses or sinuses so
placed that pressure cannot be borne from the main

stem ; it is then customary to immobilize by the double hip-splint with a longer or shorter section of the main stem on the affected side removed. The double hip-splint, which is used in all cases of hip-disease affecting both joints at the same time, and in some cases of young children when the joint-sensitiveness in single hip-mischief is extreme, consists of the chest-band already described, from which two main stems pass at a point opposite the lower angle of each scapula downward posterior to each hip-joint and down the back of each limb, and separate at the bottom by a distance of from 4 to 8 inches. The lower ends of the main stems are joined by a straight bar of iron, the inner wings of the thigh-bands are usually omitted, and we generally add

Fig. 143.—Wrench for binding the Thomas splint. Modified from Moore's triple-action ratchet drill.

a lateral wing to each side. Upon this splint the patient can be moved from bed to a couch or to a carriage with very little inconvenience or pain.

Cases of double hip-disease are not so very infrequent, occurring, perhaps in the ratio of 1 to 100 of single hip-disease. The disease rarely begins in both joints at the same time, and it occasionally develops in the second joint, while the patient is lying recumbent and protected from all traumatism during treatment of the first joint. Under these circumstances it frequently occurs that the joint last attacked recovers first, though not invariably with the greatest amount of motion. Partial or complete ankylosis of both hip-joints resulting from double

disease is not so very serious if the lumbar spine is sound
and flexible, and provided the limbs are relatively
in normal position. Patients are able to walk and to
climb stairs, to sit, and to perform most of the ordinary
movements of life fairly well. The results of double
hip-disease, treated by the Thomas double hip-splint,
appear to be somewhat
better than the results of
disease in single joints. The
nature of the affection is
such, that prolonged recumbency is necessitated, and
walking is impossible before
recovery has become nearly
complete.

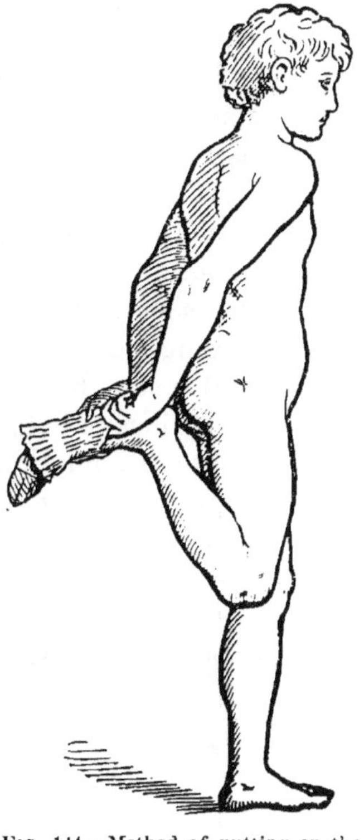

In hospital practice and
among the very poor and
ignorant, it will often be
found impossible to keep
the patients in bed with
single hip-disease as long
as we have indicated to be
desirable, and it will also be
found impossible in all
young children and in many
older ones using the Thomas
hip-splint to compel the use
of the crutches and high
patten. Parents will permit
these children to walk and
bear their weight upon the

FIG. 144.—Method of putting on the stocking when the hip is ankylosed.

diseased limb. As a matter of fact these cases do better
than we might expect. We have observed many such,
and find that some recover without flexion, rarely with
adduction, and with very little, and sometimes no
shortening. The number that have partially stiff joints

is greater than among those where treatment has been carried out in accordance with correct theories.

Now and again during the development of abscess a case will present so intense a degree of spasm of the adductor muscles that, if the patient remains fixed in the ordinary Thomas splint, knock-knee will result from adduction of the thigh, the lower portion of the leg being held by the lower part of the splint. This complication is prevented or corrected by passing a light bar of iron from the thigh band to the calf band, along the outer side of the leg, and bandaging the knee to this band as well as to the main stem.

In cases in which disease of the knee-joint appears at the same time with disease at the hip, the knee may be immobilized by joining the knee and hip-splints together, or by adding to the hip-splint a light band of iron passing down each side of the leg and around somewhat below the foot, and riveted to both the inner and outer wings of both the thigh and calf bands.

In cases of spondylitis of the lumbar region, occurring at the same time as disease at the hip, the back may be protected by a stout sling of leather, passing from one main stem to the other of the double hip-splint, or the main stem with its thigh bands and calf bands may be attached to a spinal support.

Chronic disease at the knee-joint, commonly called white swelling, or tumor albus, is the same in character as chronic disease of the hip and of the spine. Tuberculosis and syphilis in the parents predispose to its development in the child, as do the acute infectious diseases of childhood, and all those conditions which tend to a deterioration of the general health in adult life. Traumatism, however, plays a more important role in its causation than in disease at either the hip or the spine. The situation of the knee exposes it to frequent contusions, and no other joint except the ankle is more frequently subjected to sprains.

The disease commences more frequently as an osteitis than a synovitis, as is the case at the other joints; nevertheless, the relative number of cases in which the disease commences in the synovial membrane is greater here than elsewhere, and apparently this can be accounted for only by the frequency with which this articulation is subjected to injury. From bruises and wrenchings some degree of simple synovitis results; this is neglected because of the mildness of the symptoms, and in predisposed subjects it ultimately becomes the seat of tubercular infiltration. When the synovitis ceases to be simply traumatic and becomes tubercular we do not know; it appears more than probable that there is no definite time, and that it depends very much upon the constitutional peculiarities of the individual. We have observed cases in which there was no positive evidence of tuberculosis for many months after the onset of the simple inflammation, while in others tubercular synovitis commences without any remembered injury. It appears to us that, even in

186

those not predisposed to tuberculosis, a neglected simple synovitis, should it fail to spontaneously recover, may ultimately become tubercular. Tubercular synovitis, whether arising from a neglected injury or as a primary infection, is usually diffused throughout the

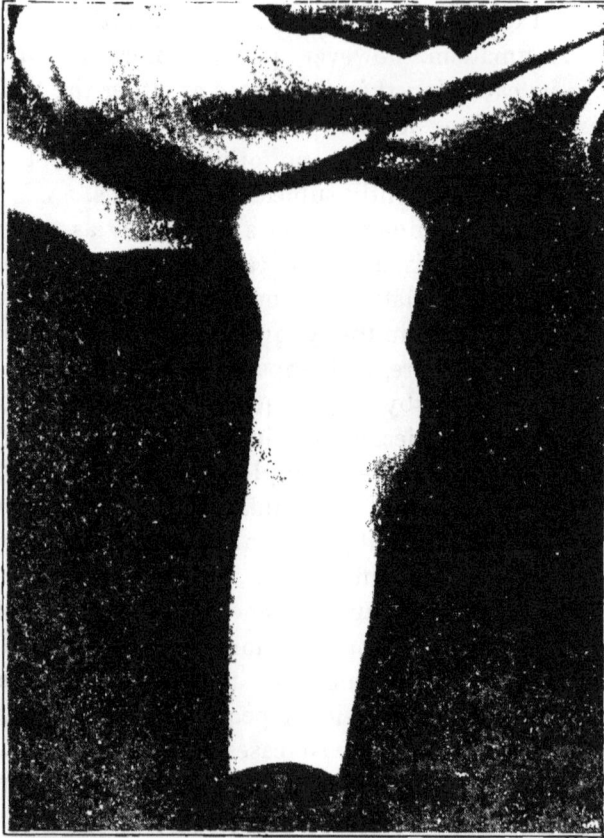

FIG. 145.—Knee-disease of four years' duration without treatment. This began as a synovitis, and only recently has shown flexion deformity and rigidity.

entire lining of the joint; only when secondary to an osseous focus have we seen it limited to a comparatively small area.

Tuberculous osteitis, here, as in other joints, may

begin as a primary or secondary focus. Its site is usually in the epiphysis, more frequently than elsewhere
in the inner condyle of the femur, next in frequency in
the head of the tibia, and least frequently of all in the
patella. The course of the osteitic tuberculosis is the
same, in a general way, as elsewhere, and the joint-
cavity usually becomes involved.

The prognosis of disease at the knee-joint may be
said to be good. The patient rarely succumbs to the
disease unless the shaft of the femur or that of the
tibia becomes involved. Un-
treated, the leg becomes flexed
on the thigh and somewhat abducted and rotated outward, the
flexion seldom exceeding 45°,
and, if ankylosed at this angle
after recovery, can be used in
walking without crutch or cane.
In severe cases, however, the tibia
may become subluxated, and the
outward rotation and knock-knee
be so great that the limb is practically useless. Tuberculous abscess
occurs somewhat less frequently
than at the hip, but when present
is no bar to a good result. Knee-

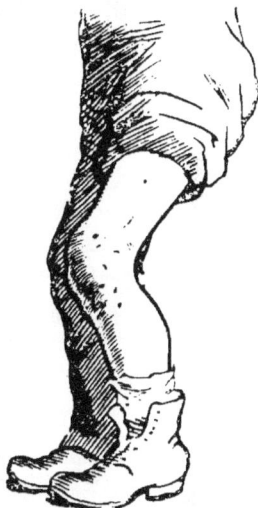

FIG. 146.—The usual appearance of a case of knee-disease.

joint disease, untreated, results, as
a rule, in ankylosis or greatly restricted motion, and
that usually with considerable deformity.

When disease at the knee is subjected to efficient
mechanical treatment the result is better as to deformity, function, and duration than is the case at any
other of the larger joints. In no case, unless the disease has been accompanied by great displacement for
a long time, should there be recovery with deformity,
and ankylosis should rarely remain; and this holds

good even when great destruction has taken place, provided there has been no subluxation backwards, or the rarer deformity of hyperextension of the tibia on the femur.

At the knee-joint, more frequently than elsewhere, do we find the difference between synovitis and osteitis clearly defined in the early stages of the disease; later on the dividing-line fades away, one condition merges into the other, and we have all of the positive symptoms of both synovitis and osteitis. At this time an excision will reveal more or less complete tubercular infiltration of all the structures composing the joint.

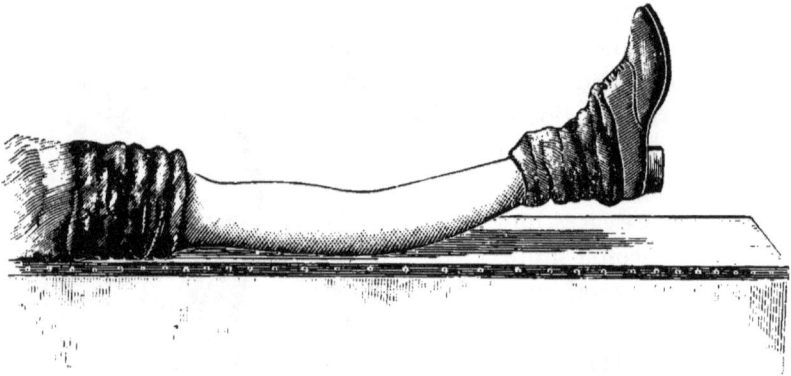

FIG. 147.—Knee-disease showing hyperextension deformity in place of the usual flexion-deformity. These cases are very rare.

It is not necessary to detail the symptoms of an acute traumatic synovitis; with such a condition we have nothing to do, but, some weeks or months later, when the acute symptoms have disappeared, there may remain, in those predisposed to tuberculous affections, and in those too impatient to give the time and attention necessary for a complete cure, a certain disability. The joint may not be found to be quite as strong as formerly, it may tire more easily, there may be slight limping after a long walk or towards the end of a day's work. Examination reveals a slight increase

of the fluid normally in the joint; the bony outlines
are less distinctly seen; the patella may or may not
float when the limb is fully extended and the synovial
sac is compressed, both above and below; there is a
slight, springy resistance to full extension; there is
usually some tenderness to pressure over the internal
lateral ligament. Rarely is there any complaint of
pain, any local elevation of temperature, or any general
tenderness. This condition may remain with little if
any change, for may months, but ultimately the bony
outlines become less and less distinct, circumferential
measurement with the tape shows a considerable increase in the size of the joint, and the part assumes all
the characteristics of a tubercular synovitis, and follows
its usual course.

A tuberculous synovitis may commence without any
remembered injury to the joint. The first disability
noticed is a slight limp after unusual fatigue, and an
examination reveals the bony outlines obscured by a
pulpy, semi-fluctuating distention of the joint. There
is no pain, or restlessness in sleep; no local heat or
tenderness to palpation; no floating of the patella or
true fluctuation; no atrophy of the muscular masses
above and below the joint; there may or may not be
increase in size of the joint by circumferential measurement, and motion is practically normal. Less frequently than in traumatic cases is there limitation to
full extension, full flexion being the first motion restricted. The restriction is springy in character, and
evidently due to the thickening of the joint-capsule, and
a very different affair from the resistance occasioned
by the involuntary muscular spasm which accompanies a tuberculous osteitis. These cases, as a rule,
progress very slowly, but sooner or later the bone is
invaded by the tuberculous growth, and the symptoms
of an osteitis are added to those of a synovitis. Fig.

145 illustrates such a case, in which swelling had been
present for four years, and where, only within a few
weeks, had there been stiffness at the joint or a tendency of the leg to become flexed, and in which as yet
there had been no pain and no tenderness. Tuberculous degeneration may be considerable, and true fluctuation within the capsule may appear; the joint may

Fig. 148.—Showing the bed-splint. Fig. 149.—The caliper splint.

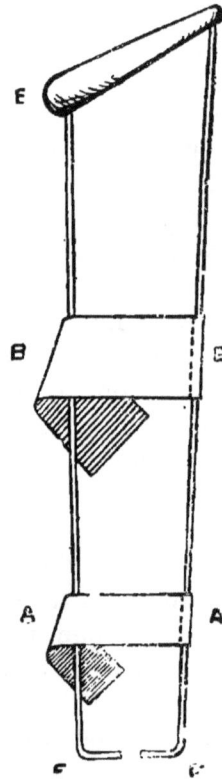

even rupture, and sinuses form, but this is rare before
the bone becomes involved. More often some shrinking of the new tissue takes place as the bone becomes
involved, and at times all swelling disappears, and the
joint presents only the characteristics of tuberculous
osteitis of the dry form.

Tuberculous osteitis invariably commences with a
limp. This may be noticed for only a short time in
the early part of the day, and days together may pass
without any limping at all. After a time the child
becomes restless in sleep and may scream out during
the first hours of the night. Rarely is there any complaint of pain until much later in the disease. Examination at this time reveals nothing abnormal in the
appearance of the joint; the bony outlines are distinct,
and no swelling can be seen or felt
anywhere. There is, as a rule, some
slight elevation of the local temperature, and there is often some bony
point distinctly tender to firm pressure. More often than otherwise
this tender point is on the inner and
lower surface of the inner condyle of
the femur. There is always present
the involuntary muscular spasm,
characteristic of tuberculous osteitis,
restricting to a greater or less extent
the normal motions at the joint.
The full degree of flexion is first
lost, and soon the leg cannot be fully
extended. Shrinking of the thigh
and of the calf-muscles comes on
early, and, with the involuntary
muscular spasm, make the only constant and characteristic diagnostic

Fig. 150.—The old form
of splint with patten
bottom no longer in
use.

symptoms. The degree of possible motion gradually
diminishes, and false ankylosis results; the leg is flexed
on the thigh to an angle from 135° to 90°, some outward rotation and abduction of the leg on the thigh
takes place, and walking becomes difficult. By this
time, and in some cases much earlier, pain is complained of and may be very severe; the condyles

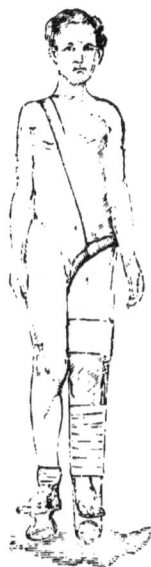

of the femur become broadened and thickened, but in
some cases the head of the tibia is the part alone involved. The synovial membrane may be invaded by
the tuberculous growth, or a tuberculous focus may
rupture into the joint, and the whole membrane become
at once infected. Now all the positive symptoms of

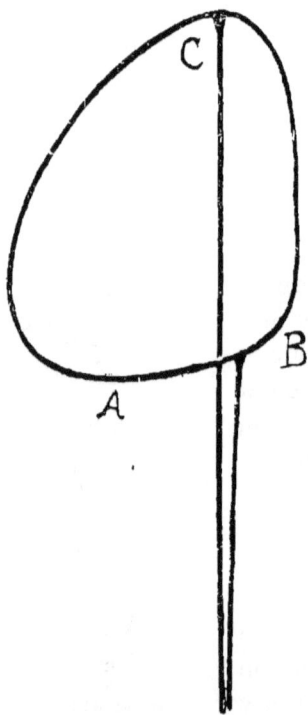

FIG. 151.—The Thomas knee-splint,
showing the inner bar, B, placed
farther to the front than the outer
bar C; A, is the lowest part of the
ring; upon this rests the tuberosity of the ischium.

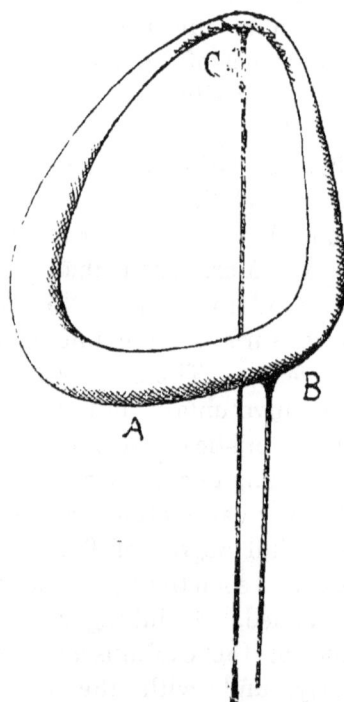

FIG. 152 —The ring of the Thomas
knee-splint after padding.

tuberculous arthritis may be said to be present. Tuberculous abscesses form in very many of the untreated
cases, and opening spontaneously, may lead into the
joint or only into bone-cavities.

In rare cases the onset is sudden, and both bone and

synovial membrane appear to be affected at the same
time. In these cases the symptoms are severe, the
progress rapid, and few joints escape rupture if not
opened by the surgeon.

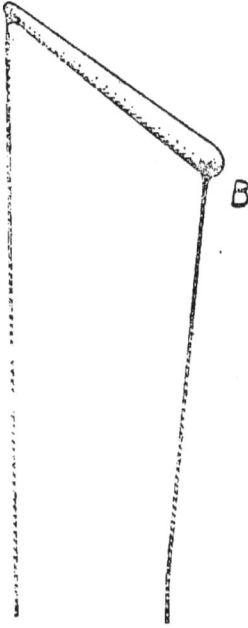

FIG. 153.—Showing the front view of the ring of the Thomas knee-splint.

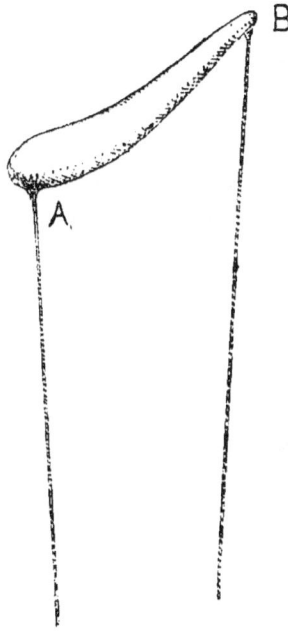

FIG. 154.—Showing the back view of the ring of the Thomas knee-splint.

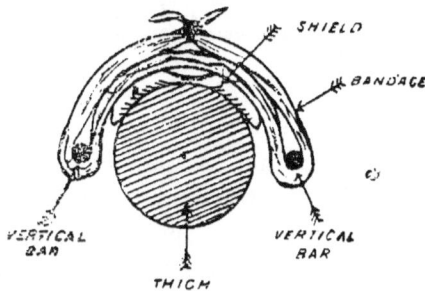

FIG. 155.—Section outline of thigh, splint, and pressure pad.

Multiple osseous foci are rarely found, but we have
observed a focus at the inner condyle (which infected
the synovial membrane by contiguity, and did not go

on to the formation of a tuberculous abscess) coincidently present with a focus in the region of the epiphyseal line on the outer side of the bone, which lead on to liquefaction and spontaneous opening.

Of the conditions which simulate tuberculous disease at the knee-joint none is so difficult to differentiate as the hysteric affection. It usually simulates the osteitic form of the disease, in which there is no change from the normal contour ; but in the shapely limb of a well-developed young woman, supplied with an abundance of subcutaneous fat, the pseudofluctuation of tuberculous synovitis in the early stage may be closely simulated.

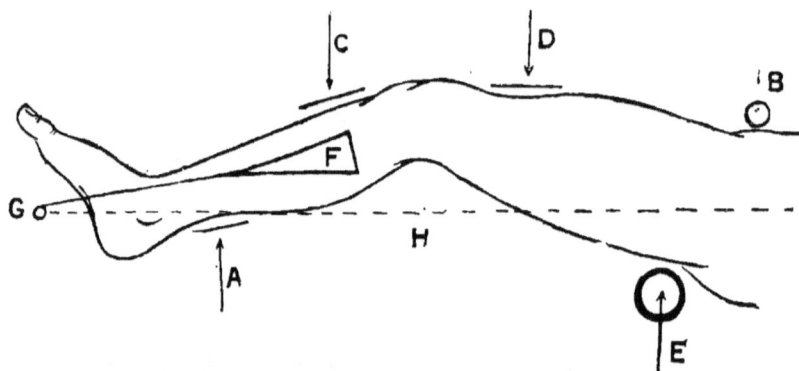

FIG. 156.—A, forward pressure of strap at the back of the ankle ; E, forward pressure of the padded ring at the upper part of the thigh ; C, backward pressure of the pad above the knee ; F, adhesive plaster for downward' traction ; B, the front of the padded ring.

In these cases we have only the presence and exaggeration of subjective symptoms to aid in making the diagnosis, and we cannot escape the knowledge that it is not impossible for true tuberculous disease to be present in an hysteric patient as well as in another. The simulation of tuberculous osteitis is even closer. The patient walks with a limp, complains of pain, the leg is somewhat flexed, the joint motions are restricted, there is tenderness to pressure and increased heat about the joint, and the circumference of thigh and calf may be less than those of the other side. Practically, all the symptoms except the tuberculous abscess may be present, and only the trained eye of the neurologist, accustomed to recognize hysteric manifestations, or the hand of the surgeon, practised to appreciate the resistance of the involuntary spasm of true bone-disease, may be able to make the diagnosis. On certainly one occasion a knee-joint has been laid open for excision and found perfectly healthy by a surgeon who disregarded the diagnosis of a neurologist and an orthopedist. As a rule, the hysteric joint is not accompanied by muscular atrophy of the thigh and calf, and

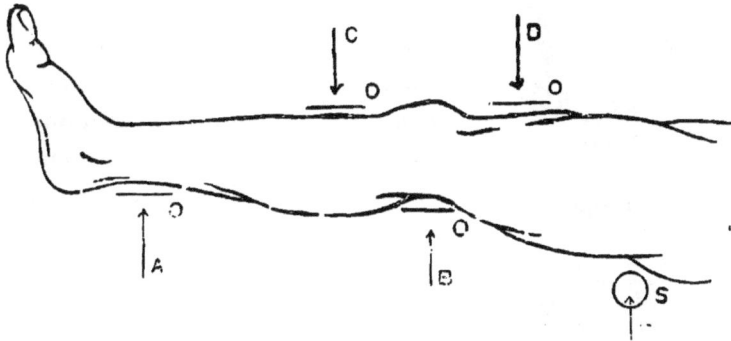

FIG. 157.—Showing the direction of supports and pressure pads.

the absence of this symptom should always be regarded as a significant fact. Muscular atrophy, however, does arise from disuse, and when present must not be taken as positively conclusive evidence in favor of a tuberculous inflammation of the bone ends.

Acute traumatic synovitis should be readily excluded on account of the history, the heat and pain, and the fluidity of the contents of the greatly distended joint-capsule.

The differentiation from rheumatic inflammation should be readily made. The suddenness of the onset and the acuteness of the symptoms are out of all proportion to that which ever occurs in tuberculosis.

The results of a gonorrheal or septic inflammation of a joint somewhat resemble certain cases of tuberculous disease, and, if the history of the case be concealed, may be confusing. In these cases, during the acute attack, the diagnosis is readily made because of its acuteness: when the acute symptoms have passed a certain amount of induration is present and is more dense to the touch than that present in tuberculous synovitis, and it is always accompanied by restricted motion at the joint. The restriction is a mechanical one resulting from the inflammation, and in no way feels like that given by involuntary muscular spasm. If there be doubt about the character of the resistance an anesthetic will settle the question. Rigidity due to muscular spasm will be lost; that due to the results of an inflammatory process will be mostly present.

A spinal arthropathy, happening to follow a traumatism, might puzzle one not accustomed to handle joints. In these cases the bone ends are increased in size

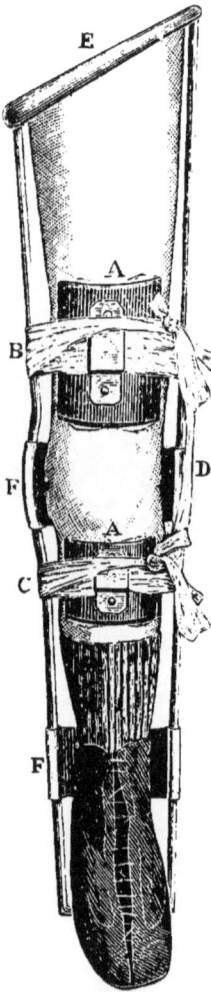

FIG. 158. The caliper splint. E. the ring around the upper part of the thigh: A, pad for backward pressure: B. bandage: C. bandage: F. leather sling for support at the back of the limb: D. a strip of bandage fastening together the pressure-pads to prevent slipping and consequent loss of pressure.

without being tender to pressure or accompanied by
pain ; the joint-distention is more fluid to the touch, and
may contain semidetached masses, firm, and of considerable size, the joint-mobility is great, soft crepitus may be
heard or felt, and the patients almost invariably present some other indication of locomotor ataxia.

As at other joints, the treatment of the disease may
be mechanical or operative ; and, with the exception of

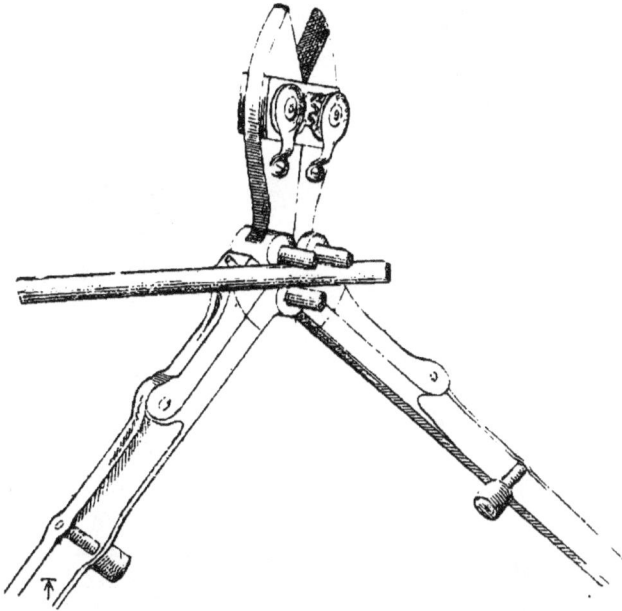

FIG. 159.—Thomas' cutters and benders for changing the bed-splint to the
caliper.

the elbow, no joint is more favorably placed to give a
rapid and satisfactory result whichever line of treatment be chosen.

The principles of treatment are the same as elsewhere, namely, complete immobilization from the
earliest possible moment until a cure has been effected,
and relief from weight-bearing until convalescence is
well established. When deformity is present it should

be rapidly corrected, since the healing process cannot
go uninterruptedly forward while the angle of flexion
is changing, and since recovery takes place more rapidly
with the limb in full extension, and with a far better

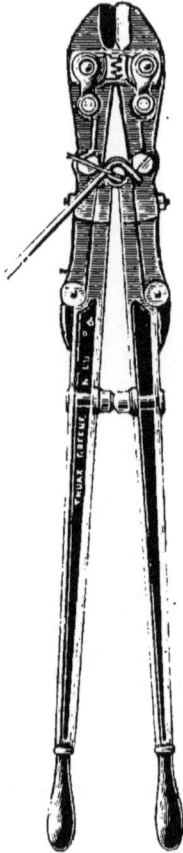

FIG. 160.--Ridlon's cutter and bender
for changing the bed-splint to the
caliper. This tool bends rods from
$\frac{3}{16}$ to $\frac{3}{8}$ in. to a right angle.

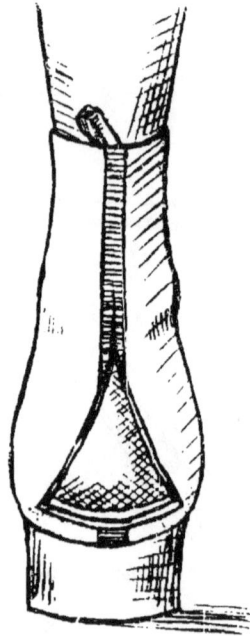

FIG. 161.—Showing the counter of the
shoe cut away to relieve pressure
on the tip of the heel.

functional result. The deformity may be best corrected
by the greatest continued-leverage force that can be
tolerated, accompanied by fixative traction : or the same
result may be accomplished by careful manual correc-

tion under an anesthetic, followed by complete immobilization.

Plaster of Paris, which has been used more extensively in disease at the knee than elsewhere, is a very

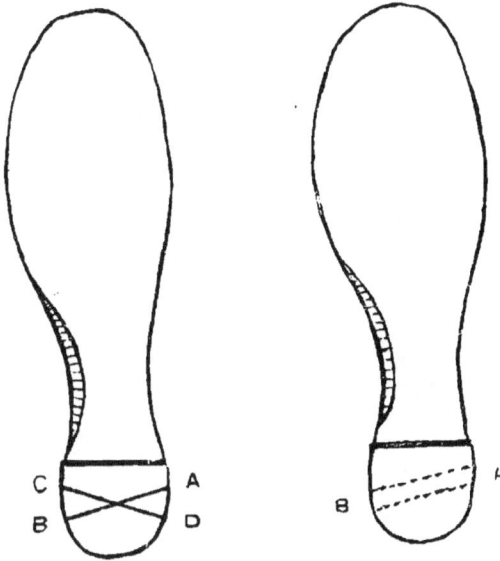

FIG. 162.—Methods of boring the heel of the shoe when the caliper splint is used.

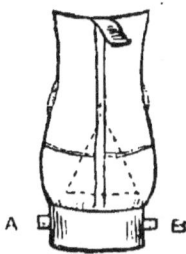

FIG. 163.—Rear view of shoe with tube in the heel for using the caliper splint.

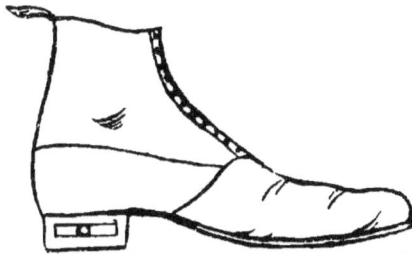

FIG. 164.—Side view of shoe with heel slotted and tube inserted for use of the caliper splint.

convenient method of treatment for the general practitioner, the general surgeon, and others unskilled in the use of mechanical devices. When used it should extend from the ankle to the perineum, and in severe

cases, where it is difficult to control the tendency of the
leg to flex on the thigh, it should extend from the toes
to the waist. All splints that are shorter than the distance from the foot to the pelvis lose enormously in

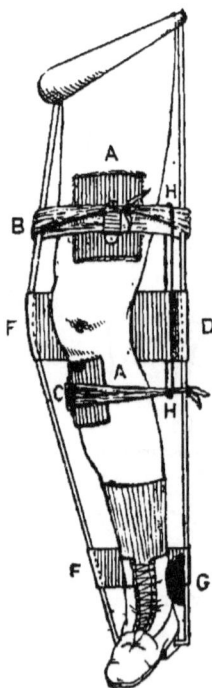

FIG. 165.—Method of applying pressure-pad to correct abduction deformity.

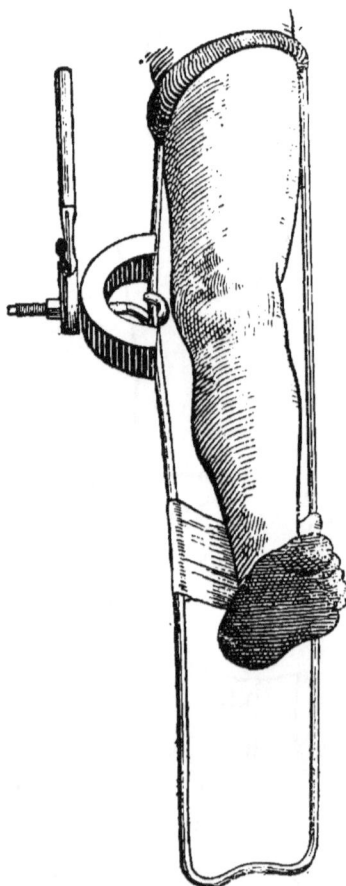

FIG. 166. Showing the tool used to
draw the inner bar away from the
swollen knee. This form of splint
is known as the bed-splint and
shows the dimple at the bottom for
tying the webbing strips from the
adhesive plasters when traction
is employed.

effective immobilizing power, and those adjusted with
innumerable screws and ratchets are expensive and of

little use. It is a pet delusion of many surgeons that interarticular pressure can be relieved by traction through adhesive plasters applied to the skin. These are applied on each side of the leg from the knee downwards for the traction force, and to the thigh from the knee upwards for the countertraction, the adjacent ends of the upper and lower plasters being apart but an inch or two. We have even seen them applied so that the upper and lower pieces overlapped, and this by an orthopedic surgeon in an orthopedic institution ; the absurdity of the arrangement, as a means of traction on the bones, being ignored and even denied when pointed out, because the patient chanced to improve somewhat under that plan of treatment. Any traction-splint to be effective must extend to the tuberosity of the ischium for its point of resistance in countertraction, and if used as a walking-splint, must extend below the foot. As a matter of fact none of the splints designed with the central idea of giving traction are so constructed.

The splint which we recommend, is known as the Thomas knee-splint, and is now used in only two forms ; the bed-splint and the caliper. The form of the splint fitted with a patten at the bottom is no longer used by us, and was not used for some years by the late Mr. Thomas in any but very exceptional cases.

The bed-splint consists of a ring of round iron to which is welded a long loop of the same material, going some inches below the foot. The ring, in shape, is an irregular ovoid flattened in front, and drawn out at the posterior and inner outline of the thigh, and, as here observed, the inner rod of the loop, B, is joined more anteriorly than the outer rod, C. The ring slopes from without inward, and from before backward in such a way that the point A, upon which rests the tuberosity of the ischium is the lowest part of the ring. The angle formed by the plane of the ring and the inner bar is about 135°, and the anterior angle formed by the anteroposterior plane of the ring and the inner bar is about 145°. The thickness of the iron depends upon the weight of the patient and is from $\frac{3}{16}$ to $\frac{3}{8}$ of an inch.

In making the ring the end should be joined by welding, and the side bars of the long loop are joined

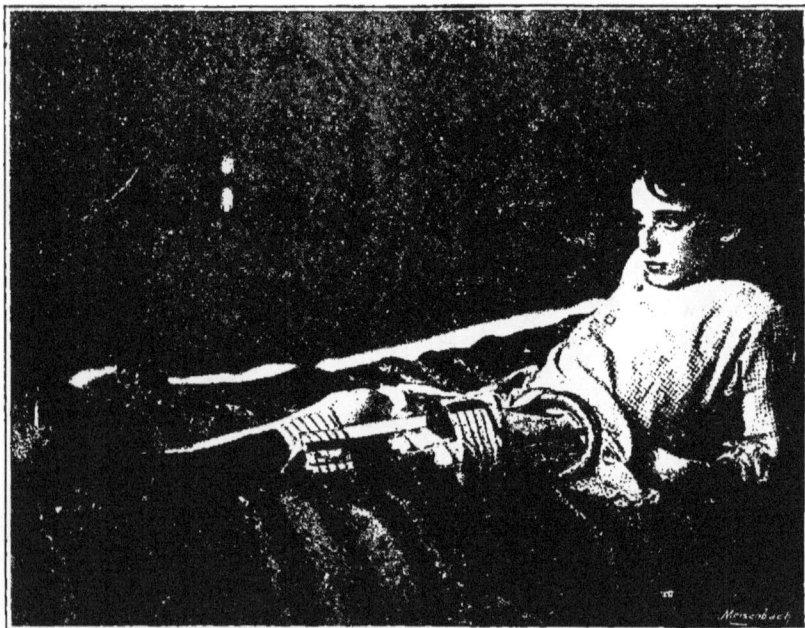

FIG. 167.- Showing the caliper splint adjusted in an old case of osteitis.

to the ring in the same manner. Few surgical instrument-makers are good blacksmiths, and therefore find it easier to braze than to weld, but a brazed joint breaks on bending, while a welded joint holds fast. The lower end of the long loop is dimpled somewhat to receive and retain the straps from the adhesive plasters. The ring is padded with boiler felting to the thickness of about half an inch on its outer portion, and from one to one and a half inches in thickness at the inner posterior portion upon which the tuberosity of the ischium is to rest, and then covered with basil leather, or tan sheepskin, put on wet, and sewed after the manner of the harnessmaker along the lower and outer border of the ring, where the seam will not chafe the patient. Two strips from three to four inches wide of the same leather are sewed by one end to one of the side bars, the other end being left free and of sufficient length to be drawn across to the opposite bar, and when sewed there to form a support for the back of the limb when the splint is applied ; one of these strips is to be placed at the back of the knee and the other at the back of the ankle.

The splint is applied by slipping the ring on over the leg and pushing it well up against the tuberosity of the ischium. If fixative traction is to be used strips of strong adhesive plaster, in width about one-fourth the circumference of the leg, and in length equal to the distance from the knee to the ankle, to the lower ends of which pieces of strong tape, webbing, or muslin bandage have been sewn, are applied to the outer and inner surfaces of the leg. If these pieces of adhesive plaster are supplied with narrow, oblique, lateral strips for winding around the leg, they will remain much longer attached to the skin. The plasters applied are held in place by an ordinary roller-bandage. The surgeon now grasps the patient's foot and pulls steadily downward, at the same time pushing the splint upward, and having straightened the limb as much as the patient will tolerate, ties the tape terminations of the adhesive plasters at the dimple at the lower end of the splint. A still better way consists in threading loops, attached to the end of the extension strips, with strings, and. after pulling, winding in spiral fashion the string on either side round the bars of the splint until they meet below. where they are tied. By this expedient all pressure on the ankle is avoided, and the tension is longer maintained.

The lower leather cross-strap is now placed at the

FIG. 168. Tuberculosis of the knee and of the elbow. Showing the caliper-splint in position and the " halter " adjusted to the arm.

back of the ankle, drawn snugly across, and sewed fast, The other leather strap is placed at the back of the knee, or at the back of some part of the thigh if the knee is too greatly flexed to rest upon it, drawn across to the opposite bar and sewed there. The knee is now pressed backward, straightening it as much as the patient will tolerate, and held there by a roller-bandage carried to and fro across the front of the limb around first one side-bar and then the other; or a thick pad may be placed across the lower end of the thigh, well down upon the patella, and backward pressure made by a strong strip of bandage passed across from side to side and somewhat downward and tied to each side-bar by a half-hitch, and then carried across the pad and tied. After this the traction-tapes at the bottom are again tightened. The limb is left thus, if everything remains in place, for two or three days, when it can again be made straighter and the fastenings made tighter. In this way the limb is straightened. If the limb is to be straightened at once, the patient being anesthetized, it is better to apply the caliper form of the splint as the one giving better fixed-traction. This will now be described.

The caliper-splint is made from the bed-splint by cutting off the lower end of the loop and bending an inch or more of side-bar inward at a right angle. The bed-splint is at first applied and pushed well up on the straightened limb, a point is marked on each side-bar half an inch below the sole and another an inch or an inch and a-half below this: the side-bars are cut off at the second point, and the bend is made at the point first marked. Fig. 160 shows the tool used in cutting off these bars, and the process of bending them. The shoe is cut at the heel, as shown in Fig. 161. This mutilation of the shoe is often necessary to prevent abrasion of the heel in walking. A hole is next bored through the heel of the shoe, or a slot is made by second hole crossing the first and a tube inserted. Into the hole, or tube, the bent ends of the side-bars are passed. the leather straps drawn fast and sewed, and the limb tied

FIG. 169.—A bad case of knee-disease.

or bandaged in place. If the knee is so swollen that the inner bar presses against it, this bar is curved with wrenches, or the tool is employed as shown in Fig. 166. When a joint has been straightened under an anesthetic, it should be left in the splint, without change of shoe, stocking, or bandages, until all pain and excessive tenderness have passed off. In a word, the joint has been more or less sprained by the maneuver and must be treated with all the consideration which a sprain demands.

The deformity corrected, the patient should be kept off his feet until the muscular spasm has subsided, when he may be allowed to walk about. If for any reason the patient has to be up before this time, he

FIG. 170.—An old case of knee-disease showing flexion, abduction and outward rotation.

should use crutches or sticks and a thickened sole under the sound limb.

The backward luxation of the tibia can be largely obviated by making the leather strip support the back of the head of the tibia, while extra backward pressure is made at the lower end of the femur. Instead of a bandage, the authors generally use two shields made of sheet iron and lined with felt, as seen in Fig. 158. This is much more effective and far simpler than the employment of a bandage.

We employ this caliper-splint very exclusively, and, in the case of children particularly, we would impress

upon surgeons the necessity of making it sufficiently
long so that the patient's heel is a good inch from the
sole of the boot. In this way the ankle escapes a jar
which were the splint shorter would surely be conveyed
to the knee.

The joint is more favorably situated than the hip for
operative interference. Pulpy masses of tuberculous
tissue may be injected with the idoform mixture,
tuberculous abscesses may be aspirated or aspirated and
injected, as may the joint-cavity itself, or any collection
of fluid may be laid open, washed out and the wound
closed without drainage. Any of these procedures, if
aseptically performed, may hasten the day of recovery :
but with either procedure septic infection may be in-

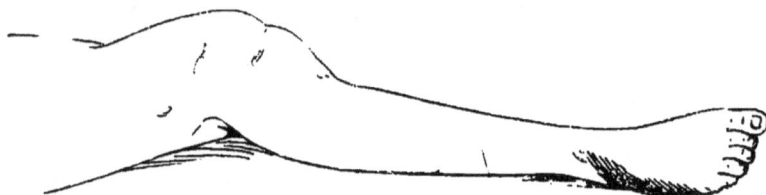

FIG. 171. The same case shown in Fig. 170 after a fair degree of correction of
the deformity.

troduced, and a serious injury inflicted. We wish to
emphasize the belief that no one of these operative
procedures is called for unless the patient is already
suffering from septic fever.

As to the major operations, erasion has no longer a
place : it has no advantages over an excision in the
final result ; motion is never regained, relapses are the
rule, and deaths from tubercular infection are frequent.
Excision as a time-saving measure in an adult case may
be employed if the patient chooses, but in our opinion
is in no other way justifiable. Any joint that can be
cured by excision can also be cured by mechanical
means without excision, and with a better ultimate

result. We, of course, refer to joints diseased and not to deformities remaining in joints no longer diseased.

The result of an incision in an adult is a stiff, peg leg with from one to two inches shortening. This may be preferred by the patient to a somewhat longer course of treatment and a movable joint, or to an amputation and an artificial limb; but the evil result of the oper-

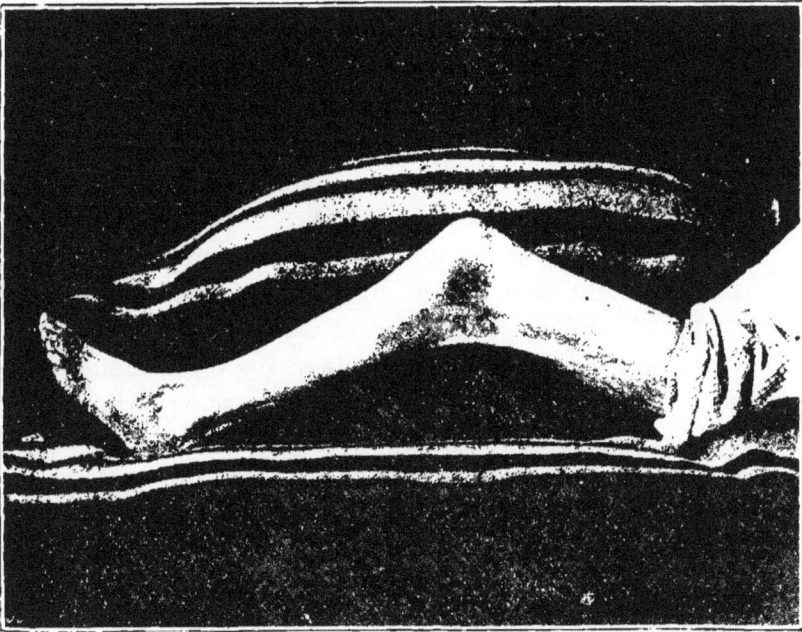

FIG. 172.—An old osteitis showing flexion deformity and false ankylosis, but no swelling.

ation in children does not end with the healing of the wound : so long as the child continues to grow, deformity of shortening increases, and when adult life has been attained the limb may be from three to nine inches short, and prove practically a useless member. In children, then, we believe that excision of the knee-joint for disease is never justifiable. Any joint that can

not be cured without excision demands an amputation.
For the correction of deformity in a cured joint several
operative procedures may be employed. In false ankylosis, brisement force, followed by complete immobilization until the part has recovered from the injury done,
is often demanded. In true ankylosis, excision, osteoclasis, or osteotomy may be demanded according to the
degree of deformity. Osteotomy is to be preferred if
the subluxation be not too great, and if the irregular
shape of the limb resulting from the operation be not
objected to. In our opinion, osteoclasis, although counted
a safer operation, is not so in these cases. Excision of
a wedge-shaped piece including the upper end of the
tibia, the lower end of the femur, and the patella with
the redundant soft parts should be chosen when the
deformity is very great, and when the neatest possible
contour of the limb is an object in the result. The
details of the operation are not demanded as they
would be for an excision when disease is present, since
the only direction necessary is to remove everything
that is in the way of complete restoration to a straight
line, and the only caution required is to remove enough
to relieve all strain posteriorly, and all pressure between the sawed ends of the bones. Clean surgery and
closure of the wounds without drainage, and complete
immobilization go without saying.

The diagnosis of a cure of disease at the knee-joint
is the same as at any other joint, namely, absence of
pain, swelling, tenderness, muscular spasm, no increased
restriction to the range of motion in joints where there
is motion, and no progressive tendency to deformity in
joints where there is no motion.

The ankle-joint is more frequently sprained than the knee: hence traumatism here plays a more important part in the etiology of chronic joint-disease than at any other joint. In all other respects disease at this joint is influenced by the same predisposing causes as elsewhere. Primary tuberculous osteitis is comparatively rare; while tuberculous synovitis is comparatively frequent.

The symptoms of ankle-joint disease when following a neglected sprain present a continuance of the symptoms due to the injury, namely, swelling about the malleoli and in front of the joint, more or less disability from restricted motion, and tenderness, and not infrequently there is pain. The foot becomes extended, walking is more difficult, the swelling increases, the lower end of the tibia and fibula become thickened, the bony outlines of the malleoli are lost in a pulpy swelling, and ultimately one or more tuberculous abscesses form at one or both sides of the joint. Tuberculous synovitis commences with a limp; soon there appears a pulpy swelling about one or both malleoli, the foot becomes extended and the normal motions are restricted, the swelling increases, and the part presents all the symptoms detailed above as developing upon a traumatic synovitis.

In the osteitic form of disease, the first symptom is the limp, pain comes on earlier and is more severe, the foot becomes extended, and all motions may be abolished before any swelling can be made out. Sooner or later the synovial membrane becomes involved, and the symptoms of a tuberculous synovitis appear; or a small

FIG. 173.—A tuberculous ankle with sinus on the outer side.

spot of pulpy swelling may appear at one or the other side of the foot without involving the joint-capsule.

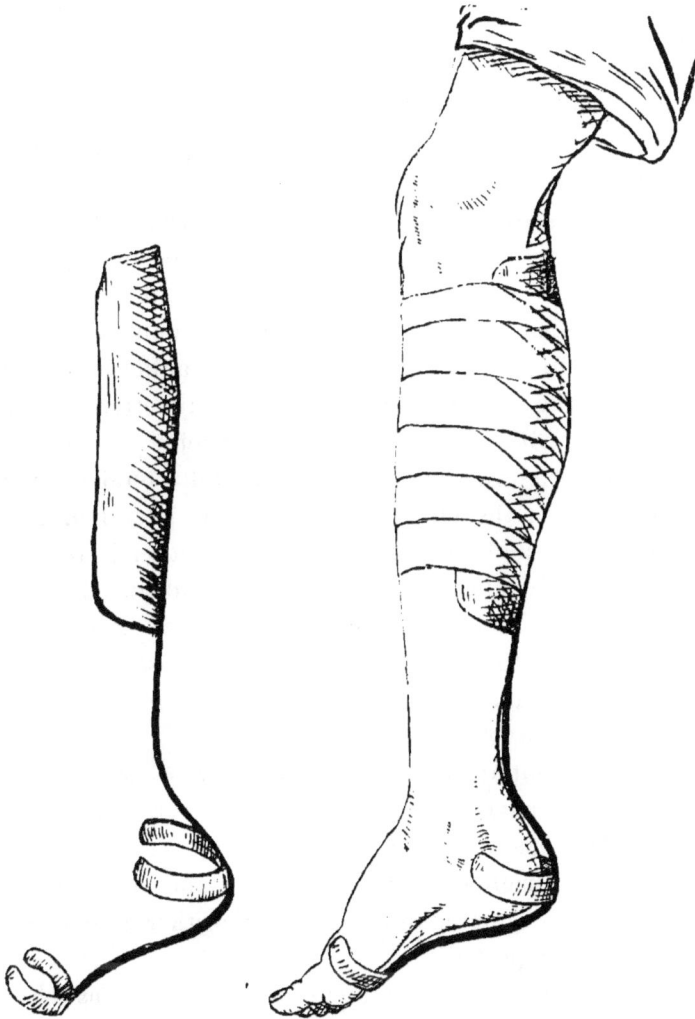

FIG. 174.—The crab splint. FIG. 175.—The crab splint applied.

As already indicated, this pulpy swelling, if left un-treated, almost invariably goes on to the formation of a tuberculous abscess, which increases up to a certain size, and opens spontaneously. The usual tuberculous discharge persists for a longer or shorter time, granules and flakes of bone come away, and finally the swelling subsides, the sinuses close, motion at the articulation returns, and, with the exception of a few scars, the joint, in children, may be as good as ever. Permanent deformity and disability at the ankle-joint are comparatively rare.

Tarsal disease arises, as a rule, in the same way, and presents the same symptoms as indicated under the osteitic form of disease at the ankle-joint when the synovial membrane escapes. The differentiation between osteitis of the bones forming the ankle-joint and those of the tarsus is made by the direction of the restricted motion. When the ankle-joint is diseased anteroposterior motion is restricted, and lateral motion is free : when the tarsus is diseased the lateral motion is restricted, and the anteroposterior motion is free to gentle manipulation. In severe and advanced cases both motions may be restricted, and it may be difficult to decide whether the rigidity is due to disease in both places, or to muscular shortening arising from the long continuance in one position.

The differential diagnosis is mainly from sprains, and of these we have already spoken. We rarely see hysteric affections of the ankle-joint, and the points of differential diagnosis from rheumatic and septic affections are the same as in disease at the other joints, and need not be repeated.

The principles of treatment are the same as elsewhere in joints of the lower extremities, and the mechanical treatment of disease at the tarsus and in the ankle-joint are the same. If the foot is extended, it should be returned as rapidly as possible to a right angle with the leg, since in this position it recovers most rapidly, and is most useful should any stiffness remain after the cure has been effected. Immobilization is maintained by the crab-splint, which in our hands is more convenient and more effective than plaster of Paris, though the plaster-splint, applied from the toes to the garter-line, serves a better purpose here than at any joint except the wrist. The crab-splint consists of a piece of sheet iron hollowed to fit the upper two-thirds of the posterior surface of the leg. To this is riveted a flat bar of iron, $\frac{5}{8} \times \frac{3}{16}$, or such size as will hold the part firmly, and it is bent to approximately follow the outline of the back of the ankle, and heel, and the middle of the sole of the foot. At the point where it passes around the bend of the heel is riveted a cross piece of iron reaching two-thirds around the ankle and of such thickness as can be bent by the hand ; and at the end of the main bar is riveted another piece of like metal long enough to nearly encircle the foot at that point. The whole may be japanned and applied over a thin layer of cotton, or it may be covered with leather without padding, and applied next to the skin. The splint is bent to grasp the foot as accurately as possible, and held in place by a strip of adhesive plaster and a roller-bandage. Young children who can be kept off their feet, and adults who can be trusted with axillary crutches, require no further appliance, but in others the knee-splint extending from 2 to 4 inches below the foot with a pattern bottom should be used. The splint is made from the bed-form of the knee-splint by cutting off the bottom of the long loop, and welding on at right angles an ovoid ring of flat bar iron $\frac{3}{8}$ inch in diameter. The greater diameter of this ring should be from side to side, for if the greater diameter be from front to back the strain in walking is thereby increased, and it may break from the side bars. The knee-splint is supported by a webbing strap from

FIG. 176.—The long protecting splint. used at times in ankle disease.

the top ring passed over the shoulder of the opposite side. The limb is loosely held in this splint by a roller-bandage at the knee.

The operative treatment of disease at the ankle and tarsus is justifiable when the disease progresses despite effective mechanical treatment and protection. A true excision of the joint should not be attempted. The tuberculous tissue alone should be removed by a cutting rather than a scraping instrument. If one is reasonably certain of having removed all the disease the wound should be dried and closed without medication or drainage. In case of a return of the necrotic process, or in cases where there is a reasonable doubt as to the removal of all diseased tissues, the wound should be left open and packed with gauze saturated with balsam of Peru. This we believe to be a better plan than the usual dressing of iodoform and rubber drainage-tube.

The diagnosis of a cure consists in the absence of all symptoms, and increase of function by use. If ankylosis results, recovery is demonstrated by the fact that the angle of deformity is a constant and not a changing one.

Disease at the metatarsophalangeal joint of the first toe is not of very common occurrence. It is usually seen in adults or adolescents, and more often than otherwise following an injury. Cases are rarely seen before the disease is considerably advanced from neglect of treatment. The toe is more or less flexed, often extremely so, and held rigid by muscular spasm, the joint is somewhat swollen, tender to pressure, and often hot

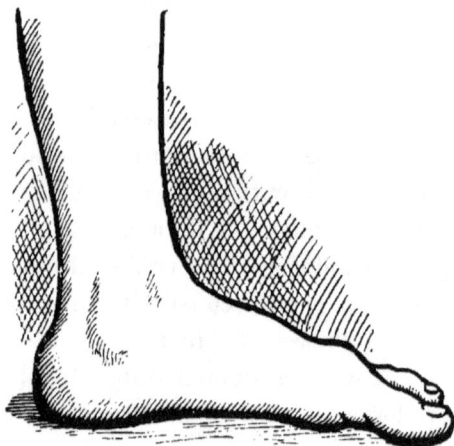

FIG. 177.—Great-toe joint disease with flexion-deformity.

to the touch. Walking is difficult, or impossible, because of the flexion of the toe and the tenderness.

The mechanical treatment consists in restoring the toe to the normal position, and permanently holding it there. This is usually most conveniently accomplished by a plaster-of-paris dressing, renewed every week or two until the deformity is corrected. During this time crutches must be used. Afterwards immobilization may

be had by a rigid metal insole, supplemented by a
block of wood an inch thick fastened to the sole of the
shoe and extending from the ball of the foot to the back
of the heel. Or a strip of steel may be laid into the
sole of the shoe reaching from the tip of the toe to the

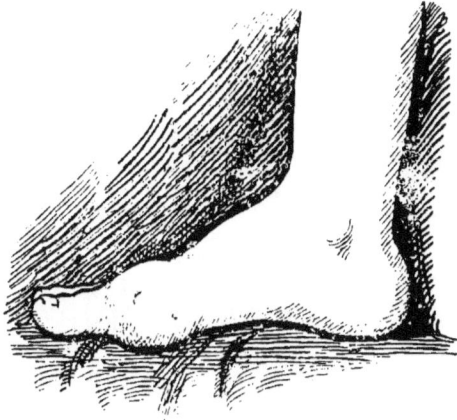

FIG. 178.—Great-toe joint disease with much swelling and sinuses.

FIG. 179.— Shoe with block on sole and kid patch over joint

heel, and screwed fast. This must be rigid enough to
prevent all bending of the sole of the shoe. If the joint
is swollen and tender the leather of the shoe-upper
should be cut away and a slack piece of kid inserted.

The operative treatment consists in an erasion of the

diseased tissues as recommended in tarsal and ankle disease. The results of treatment at the great toe-joint are comparatively rapid, and in all cases that have been under our observation full normal motion has been gained.

SHOULDER DISEASE.

Chronic disease at the shoulder-joint is not of very frequent occurrence in either children or in those adults who, from the nature of their occupation, do not subject this joint to injury; but in certain classes of laboring men, who frequently subject their shoulders to strains, wrenchings, and contusions, and who from the necessity of unceasing toil are unable to await complete

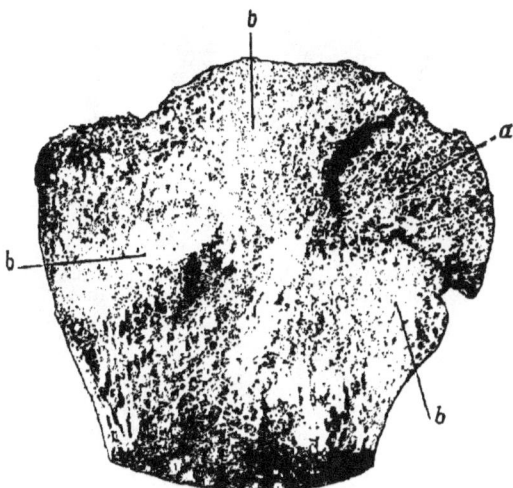

FIG. 180.—Head of humerus, sawed through in frontal plane; (a) wedge-shaped, completely separated sequestrum extending to denuded joint-surface; (bb) extensive cheesy infiltration of the head; secondary infection from the sequestrum. (From Krause.)

recovery from the injury, disease at this joint is more frequently found than elsewhere. Of these traumatic joints we shall speak later.

When chronic disease at the shoulder-joint arises without any remembered injury, it has the same etiology as tuberculous joint disease in the lower extremity, which has already been discussed. It usually

commences as an osteitis : rarely as a synovitis. The
first focus in the bone is usually in the head of the

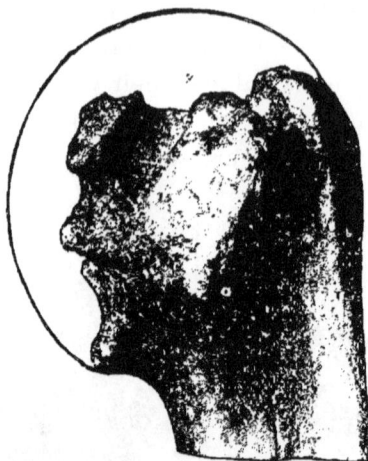

FIG. 181.—Remains of the head of the left humerus in caries. Deep erosions.
Contour of head indicated. (From Krause.)

FIG. 182.—Upper end of the humerus in a case of dry caries. Head entirely
gone, covered by a thin layer of firm tuberculous granulation-tissue. (From
Krause.)

humerus ; rarely in the glenoid cavity. Wherever the
focus may begin it gradually extends until all parts of

the joints are involved. If the progress of the disease
be rapid, liquefaction of the tuberculous material leads
to the appearance of tubercular abscess ; if slow, the
patient is said to be suffering from the dry form of the
disease, and no abscess appears.   In our experience
tuberculous disease of the shoulder-joint is more frequently dry than otherwise.

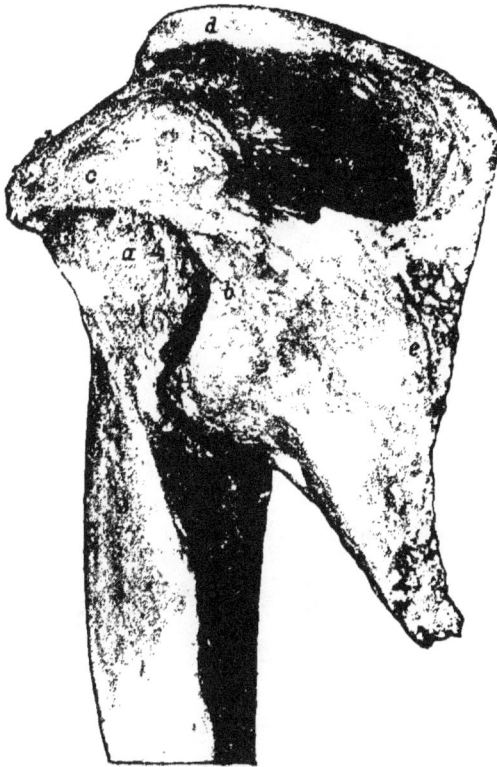

FIG. 183.—Dry caries of right humerus from front. Ankylosis. (a) Remains of
head ; (e) coracoid : (d) acromion : (e) body of scapula sawed off. (Krause.)

The traumatic form of chronic disease at the shoulder-
joint, which is associated with adult life, has scarcely
ever been referred to by writers. The chronic symptoms gradually become engrafted upon a slight syno-

vitis and present characteristics quite different from tuberculous osteitis, rheumatism, or osteoarthritis. The course is much more benign than tuberculosis ; it occurs in subjects who are strong and active : it rarely, if ever, results in suppuration : it invariably yields to treatment with but little variation in the length of time. From chronic rheumatism it may readily be recognized : it

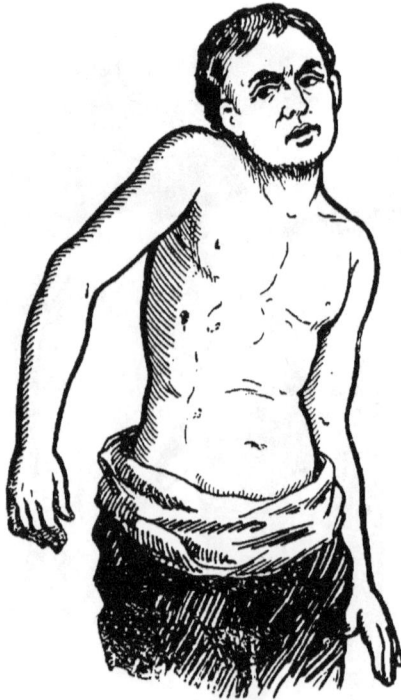

FIG. 184.—Cannot abduct the arm without pain.

results from an injury, but does not immediately follow the acute symptoms ; it is always worse after exercise : and is not accompanied by general rheumatism. With osteoarthritis it should rarely be confounded ; it never results in destruction or eburnation of joint-surfaces ; rest never tends to stiffen, but always loosens the joint ;

it presents none of the characteristic symptoms of osteoarthritis, and its clinical course has nothing in common
with it. The patient may complain fitfully, often in
accord with atmospheric conditions; but it should be
remembered that patients with fractures, injured muscles, and wounds often complain more in threatening
weather. The arm is held closely to the side; pain is
complained of over the joint and often at the insertion

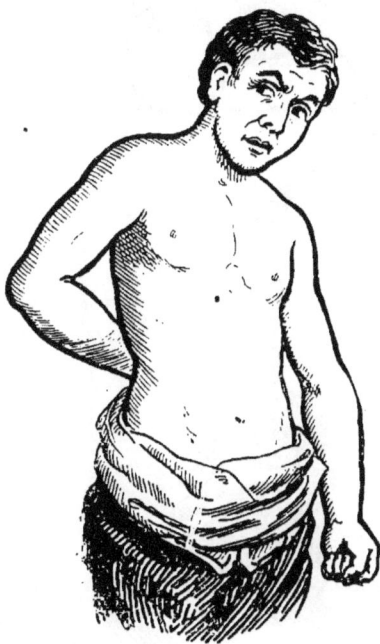

FIG. 185.—Pain in trying to put the arm behind the back.

of the deltoid muscle; the patient is unable to sleep
upon the shoulder; and should he accidentally lie upon
it he awakens with pain; acute pain is experienced
upon pressing a point just below and to the outside of
coracoid process which corresponds to the front of the
humerus. There is an inability to rotate the joint either

inwards or outwards without discomfort. Abduction is
limited, and elevation of the arm impossible. The surgeon, in short, elicits from the patient the following
story: "My arm is weak and painful; I dare not lie
upon it at night, or the pain becomes incessant; I cannot lift it from my side beyond a short distance; it

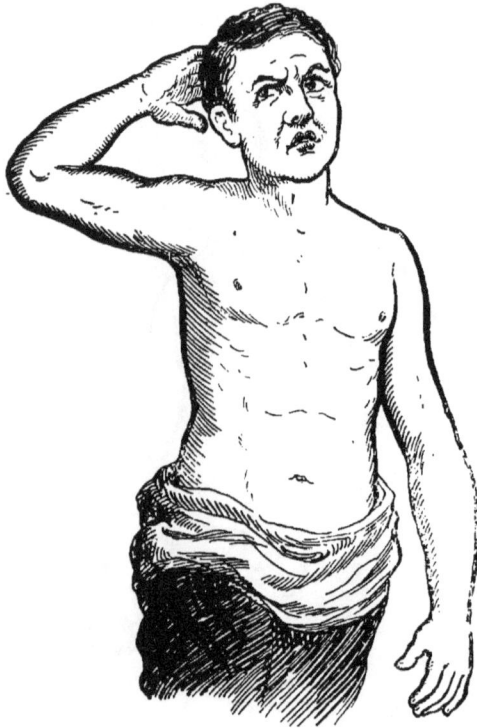

Fig. 186.—Painful when trying to place the hand behind the head.

pains dreadfully if I attempt either to get my hand
behind my back or behind my head, and yet I can
touch my opposite shoulder with no discomfort and I
can with equal ease push my arm forward." So correct
is this clinical description that one need never be at
loss to anticipate a patient's story.

Subsequently, shrinking of the shoulder-muscles takes place, and finally the limb may become wholly useless. If abscess forms, which certainly rarely ever occurs in this adult variety, the first true fluctuation will be made out in front of the joint.

In tuberculous osteitis the first symptom is usually restriction to full rotation and to carrying the arm back-

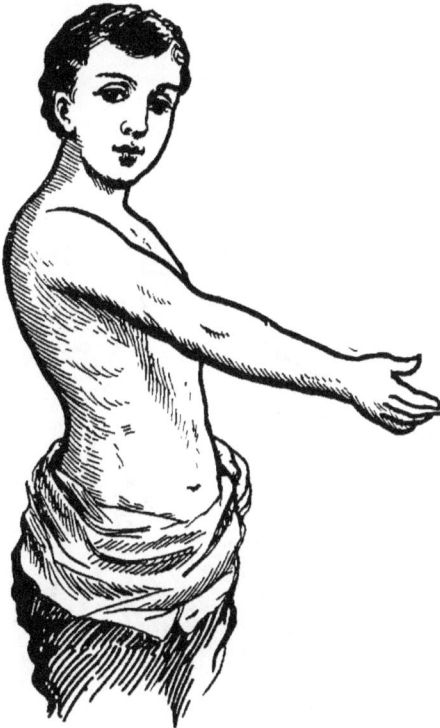

Fig. 187.—Can easily push the arm forward.

ward. The patient finds that he cannot when dressing reach behind him and adjust his clothing with accustomed freedom and dexterity, and the arm is held hugged to the side. Shrinking of the muscles comes on early and is a constant symptom; tenderness to pressure may be found over the outer anterior aspect

of the head of the humerus, and pain may be complained of at the insertion of the deltoid. The progress
of the disease, unless aided by art, is from bad to worse.
The complications are abscess and subluxation. The
tendency of the tuberculous abscess is the same as elsewhere. If untreated, it descends as it approaches the

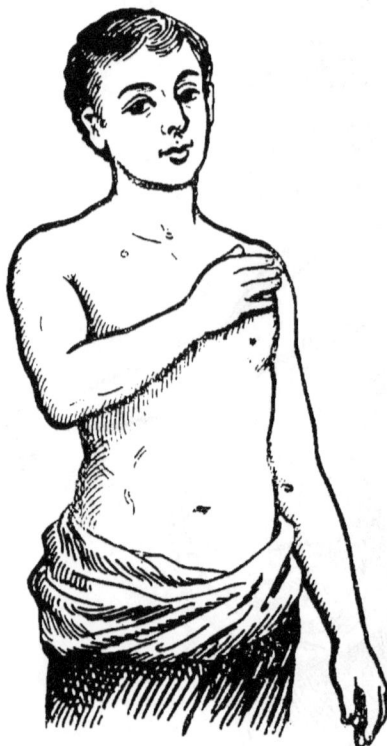

FIG. 188.—Can easily touch the other shoulder.

surface and ultimately ruptures; if treated, a large per-
centage gradually disappear without going on to spontaneous rupture. Chronic abscess in the shoulder-joint
escapes through the least protected parts of the capsule,
and is subsequently directed in its course by structures
which form a large shield over the joint. The joint is

supported and protected by two successive muscular
coats, the first or deeper being intimately associated
with the fibers of the capsule, the superficial protection
being the deltoid muscle. The muscles touching the
capsule consist of the subscapularis, long head of triceps,
supraspinatus, infraspinatus, and teres minor, the three
last being intimately blended into a continuous mass.

FIG. 189.—Showing halter applied ; also beginning of
apparent subcoracoid luxation.

In this coat weak spots exist in two places, above and
below the subscapularis. In the part above the capsule
it is supplemented within by the long tendon of the
biceps, which courses in the interval and escapes
through a hole in the capsule, in the part below the
capsule is thinnest and least protected. Pus having
escaped through one of these undefended spaces finds

itself either under the deltoid, and is forced to point near the front or back of its insertion, or in the axilla.

No attention has been drawn by previous writers to the pathologic displacement which occurs in a large number of tuberculous shoulders. It is not necessarily

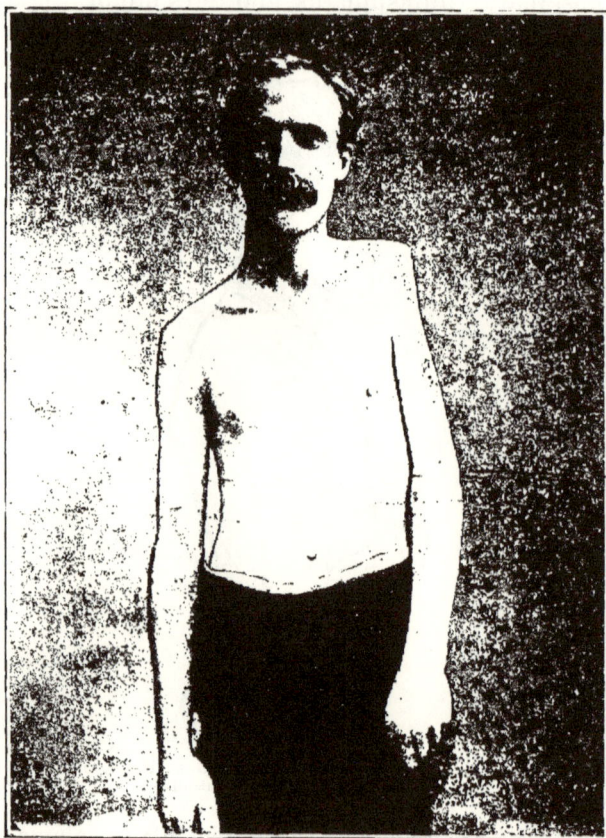

FIG. 190.—Arthritis with ankylosis. Pathologic displacement of head of humerus.

dependent upon liquefaction of the head of the humerus. It is almost always accompanied by ankylosis. This condition is not dissimilar in appearance to an old unreduced traumatic dislocation. The head of the

bone is not in reality very much out of position, but
the capsule of the joint has become shortened, and this,
in conjunction with the hugging of the arm and the
marked atrophy of the deltoid, produces a strong like-
ness to a subclavicular or to a subcoracoid luxation.

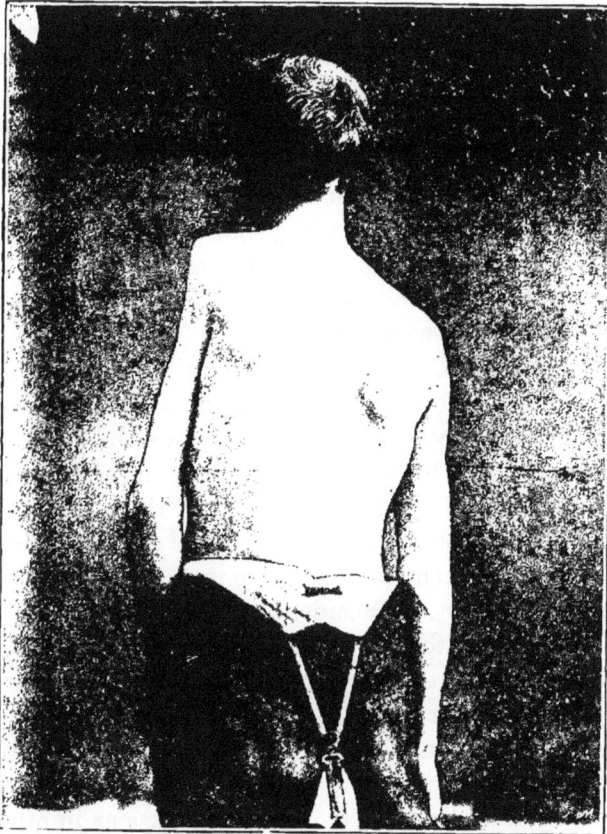

FIG. 191.—Rear view of patient shown in Fig. 190.

One of the authors (R. J.) has, on frequent occasions,
been asked to reduce such shoulders, the patients and
often their medical advisers failing to recognize the true
nature of the lesions. We will briefly relate the his-

tory of a case by way of example, which shows the danger of not making a correct diagnosis. A gentleman brought his son, a youth of 19 years, to consult about his shoulder which, it was said, had been dislocated for over two years. He had already been to the metropolis and had consulted three surgeons of repute. One had suggested excision of the head of the humerus, another manipulation under anesthetics, and the third counseled a let-alone policy. All of them, however, acquiesced in the belief that they were dealing with a case of neglected dislocation. When the patient was examined we had no difficulty, even before inquiry into the history of the lesion, in deciding that the case was one of pathologic subluxation following arthritis. The patient's father became convinced of the correctness of this later view when we were able, without being prompted, to tell him the main outline of the course the diseased shoulder had gone through. The facts were briefly these : The boy fell from the box of a brougham upon his shoulder ; it was painful and swollen. The family-doctor called next day, prescribed rest and evaporating lotions, and after a week's attendance left his patient with a few general instructions. In a fortnight after the accident he was able to use his shoulder, but with some discomfort. He always complained of pain on lying upon the joint. Examined two months later by an experienced surgeon, dislocation was pronounced, to the discredit and humiliation of the local practitioner, who fully believed he had made a serious initial error. In this case as in others, the differential diagnosis from an old traumatic displacement was not difficult. Ankylosis was complete, which is rarely the fact in neglected luxation.

The muscular shrinking was more marked, the head of the bone less really displaced, and the coracoid more prominent than would have been the case in an old

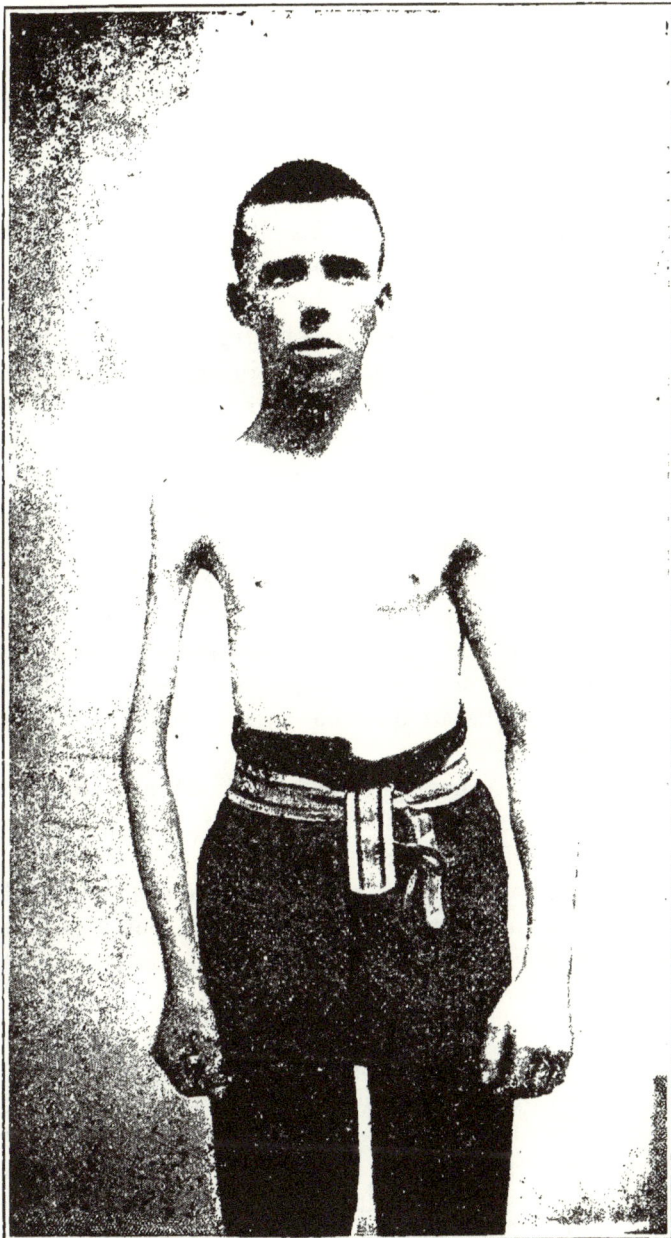

FIG. 192.—Showing muscular atrophy in shoulder-joint disease.

dislocation. How important, therefore, it is, both for the sake of the patient and the previous attendant, that the differences and the similarities between the two conditions should be noted; for it must be emphasized that tuberculous arthritis of the shoulder without suppuration, often results in ankylosis, such ankylosis being an accompaniment of the pathologic displacement.

The differential diagnosis of traumatic cases is materially aided by the history of the injury, and, although rheumatism may attack an injured joint, too much credit must not be given to the rheumatic theory. Tubercular osteitis should not be mistaken for rheumatism because of the insidious onset and absence of early involvement of the soft parts.

When shrinking of the muscles is extensive, and in cases that have experienced no pain, the part in appearance resembles certain spinal and traumatic palsies. In paralysis, however, there is abnormal passive mobility, while in joint-disease the mobility is diminished or altogether wanting.

Without treatment the disease almost invariably results in a shrunken, shortened, stiff and somewhat useless member. Under treatment, the shrinking, shortening and stiffness are lessened, and if commenced sufficiently early and faithfully carried out, a perfect cure not infrequently results with full restoration of normal function.

The mechanical treatment of shoulder-joint disease is not by any means theoretically perfect, on account of the continued movement of the thorax and the shoulder girdle during respiration, though practically the results are good. Immobilization by leverage is not easily made; neither can the joint be satisfactorily held by any of the plastic dressings which surround the arm and chest. Traction can be applied

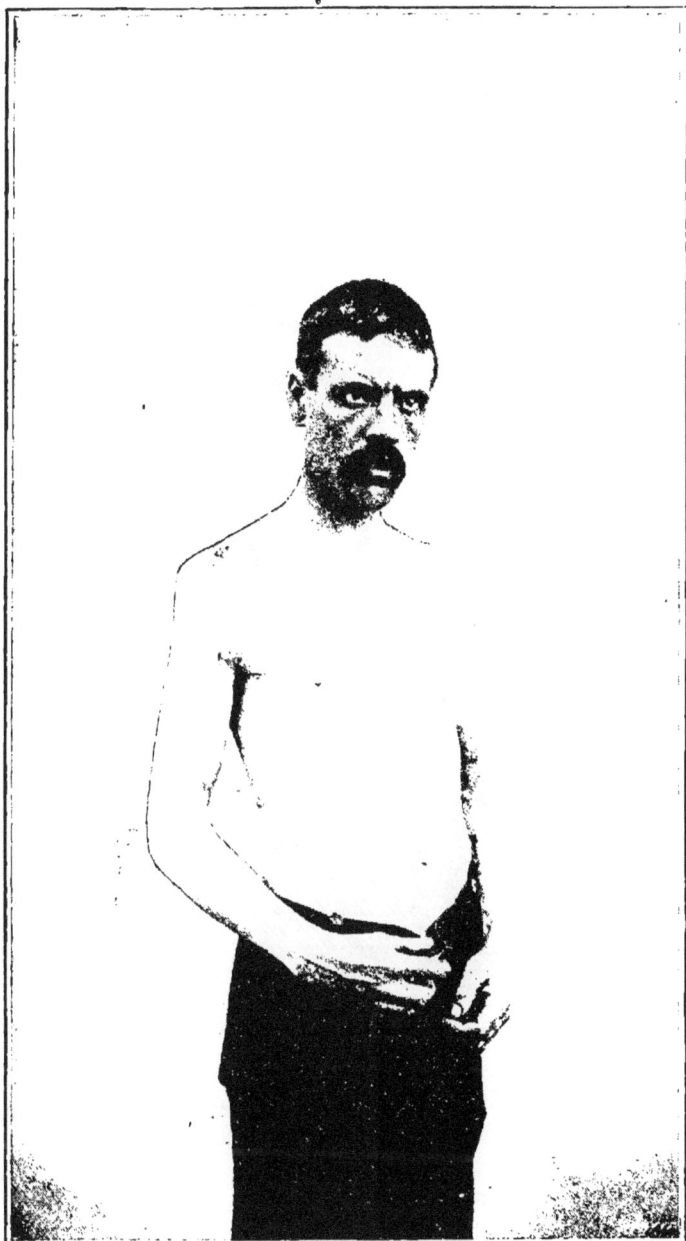

FIG. 193.—Showing subclavicular luxation in shoulder-joint disease, accompanied by ankylosis.

more readily and perfectly than at any other joint,
but its use aggravates the pain in most cases, and the
treatment by traction gives less good results than at
joints where it can be applied with less theoretical
advantage.

The treatment which we have found most useful con-

Fig. 194.—Showing halter and strapping for the treatment of
shoulder-joint disease.

sists in restraining voluntary movements by applying
the "halter" about the wrist and neck, and at times a
supplementary band from the elbow around the body.
A restraining influence is brought to bear locally by
covering the shoulder with numerous strips of pitch
plaster spread on thick paper. These strips are laid on

one over the other, covering the shoulder and extending
well down over the scapula on the back, and over an
equal distance in front.

In a few cases the weight of the arm gives rise to a
dragging pain, and it is of advantage to support the
member by a broad bandage passing from under the
elbow over the shoulder of the opposite side.

In shoulder-joint disease tuberculous abscesses can
more readily than elsewhere be aspirated, and their
cavities and fungoid masses injected with the iodoform
mixture or other medicament, but, as elsewhere, this line
of treatment has not been followed with good results
except in an occasional case.

The treatment of the adult variety of chronic shoulder-
disease is the same as if it were tuberculous, namely, rest
from pressure, friction and function. Treatment usually
lasts from 3 to 5 months ; and the test of recovery most
convenient for the practitioner is the patient's ability to
sleep painlessly upon his joint. This form never results
in ankylosis unless neglected. The test of recovery in
tuberculous cases is in a fixed and painless ankylosis,
or where ankylosis does not exist, a range of motion
which may tend to increase, but never to diminish. With
scapular movement always available, excision of the
shoulder is never called for.

ELBOW DISEASE.

Chronic disease at the elbow-joint arises from neglected injury and from constitutional dyscrasia in the same way as does chronic disease at the other articulations. The joint is fairly frequently subject to injuries so slight that they are neglected and allowed to become chronic; but here, more frequently than elsewhere, do we find chronic disease following severe injuries. As a rule, joints suffering from dislocation or fracture are immobilized after reduction until a cure has been effected; but the elbow forms an exception to this rule. The joint is so favorably situated for "surgical manipulations" and "passive motion" that few surgeons have been able to refrain from these most harmful pastimes, and as a result one is asked to treat a tuberculous elbow that would never have been tuberculous had it escaped such treatment for the original injury. We would not say that a dislocation at the elbow or a fracture into the joint should not be treated, but we believe that it would be better to leave such untreated than to commence forcible bendings before a complete subsidence of the inflammatory process. Tuberculosis may here, as elsewhere, commence as a primary or a secondary focus in the bone or in the synovial membrane; and as a synovitis it commences here next in frequency to the same condition at the knee. Chronic disease, the result of inherited syphilis, is not infrequently met with at this joint.

Disease following a neglected slight traumatism has its symptoms grafted on to those of the original injury. When first seen there is some obliteration of the normal bony outlines; thickening and some tenderness can be made out between the olecranon and the condyles on either side, and there is some restriction to full flexion and extension, but no true involuntary muscular spasm.

As the case progresses the swelling increases, the thickening becomes more pulpy and ultimately fluctuates, and sooner or later the bone is invaded, and then true muscular spasm restricts the joint motion.

FIG. 195.—The halter applied in elbow-joint disease.

A tuberculous synovitis presents no symptoms except swelling and a slight elastic restriction to full flexion and extension,'until the bone is invaded or an abscess forms.

Those cases which follow severe injuries and result from the efforts of the surgeon in the direction of passive motion follow much the same course. The thickening, the tenderness, and the restricted motion are still present when the passive motion is commenced, and they go on increasing until all the structures are involved and muscular spasm absolutely locks the joint.

The symptoms of tuberculous osteitis are restriction to normal motion, a sense of weakness often accompanied by an aching pain, and shrinking of the muscles both above and below the joint. When the tubercular focus has made its way out of the bone, whether into the joint or external to it, swelling appears, at first pulpy to the touch, and afterwards breaking down into a tubercular abscess, which presents in the majority of cases, below and to the outer side of the external condyle. The forearm becomes flexed from 40° to 70°, and motion is almost, or quite, lost. The bone-ends become thickened, and numerous sinuses appear, or in the dry form of disease little or no swelling and no abscesses may be present.

The prognosis of elbow-joint disease without treatment is not often determined, for the surgeon is greatly prone to excise this joint. The results of excision are rather better here than at any other joint, but so also are the results of mechanical treatment without operative interference. The ultimate result as to motion when treated mechanically depends of course upon the extent of the disease; a few joints recover ankylosed, but the vast majority regain normal motion when treated by efficient immobilization.

The treatment consists in placing the member as rapidly as possible in such position as will prove most serviceable in case partial or complete ankylosis should result, and retaining it uninterruptedly in that position until a cure has been effected. The position of election is that of complete flexion; such a position as will enable the person to feed himself and dress his hair

should ankylosis result. If the forearm can be bent
that far at once without considerable suffering, this is
done ; if not, it is carried as far as possible towards the
neck, and the head is bent as far forward as possible to
meet it, and the halter is snugly applied. By the end
of 2 or 3 days the head will have become erect and the
forearm will have by that much approached the desired
angle : the head is again bent forward and the slack in
the halter taken up. This is repeated until the desired
position has been gained. The halter is then left at
that length until the joint is cured. We are accustomed
to fix the wrist to the neck in full flexion the moment the
elbow is found to be effected. If this is done complete
rest is secured for the joint, as the halter prevents extension and the neck prevents flexion. If ankylosis
threatens there will be ample time to drop the wrist to
the most useful angle, and by this time the graver symptoms will have abated. The useful angle in the case
of a laboring man is about 90° ; in the case of a shop
assistant, about 100° ; while many folks with ample
means prefer an ankylosis at about 135°, which enables
them to be less awkward at table.

The halter consists of a piece of broad bandage sufficiently long to go around the wrist twice or thrice and
thence around the neck. It may be fastened around
the wrist with a " half hitch " or passed twice around
and knotted. It must be loose enough not to constrict,
and yet so tight that the hand cannot be drawn out of it.
Near the neck the bandage is again knotted, and then
tied around the neck. Sometimes the halter is left
around the neck without interference for some months.
In such cases if the surgeon can manage it the bandage
should be passed through a tube of leather which acts
as an excellent padding against the neck, and then
through a leather band stitched around the wrist.

All knots should be sealed to render it certain that

the dressing has not been tampered with. The dressing should be changed only when cleanliness demands it. Splints in the case of the elbow should be avoided; they interfere by pressure on the vessels of the arm, check reparative processes, and materially increase the patient's discomfort.

Operative procedures consist in aspiration of abscesses, injection of medicinal mixtures and excision. For the aspiration and injection treatment the elbow-joint is favorably placed, but there is no evidence to show that the ultimate cure is hastened by these measures in any but a few exceptional cases. Excision is more favored by general surgeons here than elsewhere because of the better functional results obtained ; nevertheless, we believe that any joint that can be cured by an excision will get well without an excision and results in a far more useful arm. The most imperfect results obtained by mechanical treatment give a stronger arm than the best results from an excision. Only as a life-saving measure should excision be considered, and in such a circumstance the choice must lie between it and an amputation.

The test of recovery in this joint is easily applied. It is founded on the principle that as disease advances motions become less. When the swelling has subsided and the patient feels his arm to be well, the experiment of testing is undertaken. The halter is lengthened two inches and the wrist allowed to drop : if in a few days the patient is able to lift his hand to the point from which it was released, recovery is assured and the wrist may be allowed a further drop. If ankylosis has resulted, the test of its soundness may be made in accordance with the principle that a stiff joint is free from disease when it does not change its angle by use.

Chronic disease at the wrist-joint is of rare occurrence in children, except among the children of syphilitics. In adults it more frequently than otherwise appears to follow sprains and other slight but neglected injuries. In such cases it appears as a synovitis, and the prominent symptom is the swelling of the part. Cases are met, however, in which there is no swelling, and in which the restricted motion and muscular spasm point to an osteitic focus. In these cases, as in the tarsus,

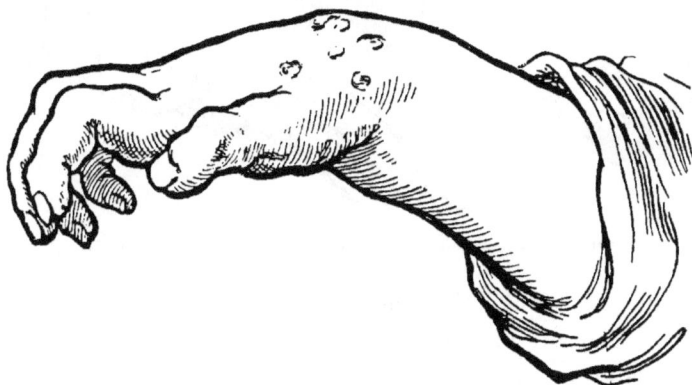

FIG. 196.—A tuberculous wrist with flexion-deformity and sinuses.

swelling, as a rule, comes on early, and the part presents heat, pain, restricted motion, and some deformity in the direction of flexion ; the patient usually presenting himself supporting the diseased part with the other hand. Left untreated all motion is completely lost, and the hand may become flexed to a right-angle to the forearm. Abscesses form, as a rule, and are generally met on the dorsum of the wrist and hand.

The prognosis for mechanical treatment is good, both

as to the position and usefulness. The treatment can be
most satisfactorily conducted by a simple splint of
sheet-iron, bent to a hollow to fit the forearm and with
a T-shaped ending at the hand. The part of the splint
upon which the hand rests should be placed at an

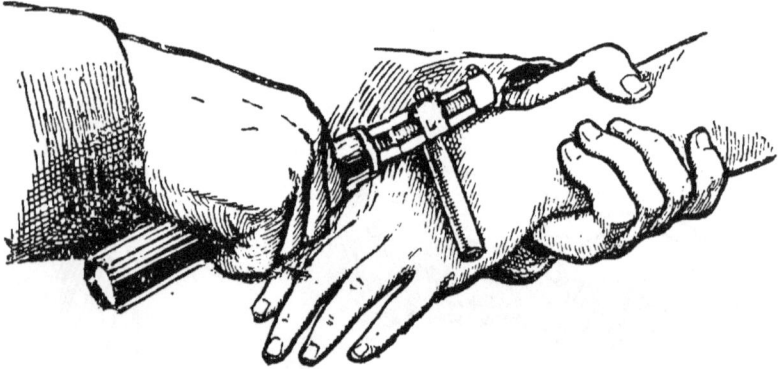

FIG. 197.—Using the wrench to break up an ankylosed wrist

angle of about 140°, and the lateral wings should be
bent around and grasp the hand at the metacarpo-
phalangeal articulation, the whole being held in place
by strips of adhesive plaster. The flexion should be

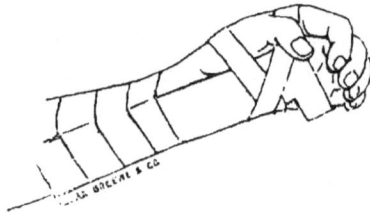

FIG. 198. A simple splint of sheet metal for the treatment
of wrist-joint disease.

overcome as rapidly as possible, and the hand immo-
blized in an extended position. This extended position is of great importance in the treatment of the wrist
and is not referred to by any writers on the subject.
If one wishes to grip powerfully, the wrist is first in-

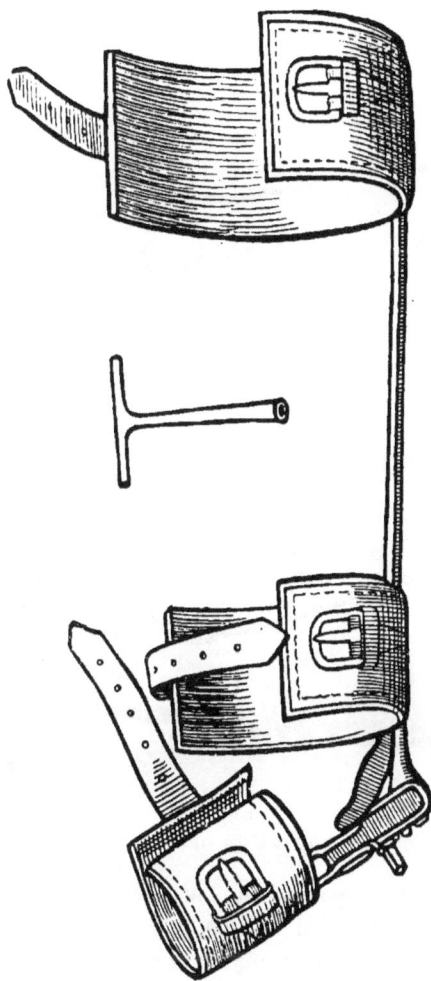

FIG. 199. Jones' adjustable splint for the treatment of wrist-joint disease.

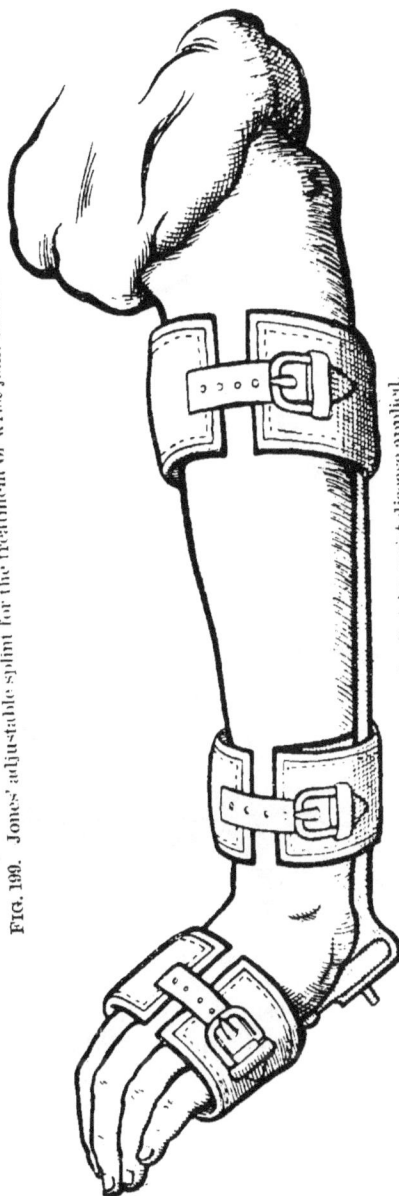

FIG. 200. Jones' splint for wrist-disease applied

stinctively extended. One cannot grip effectively with
the wrist flexed. The lesion is obvious. If disease at
the wrist results in ankylosis in the extended position
every advantage will accrue for subsequent action of
the extensors of the hand. So true is this that when a
patient consults us complaining of faulty use of the
fingers, should the wrist be ankylosed in the flexed

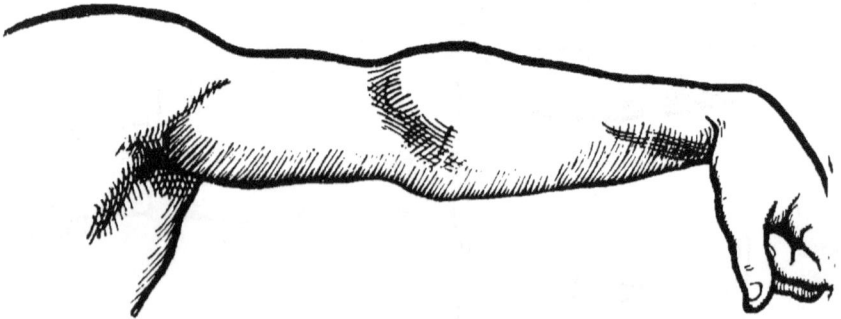

Fig. 201.—The position of the wrist for a weak grip.

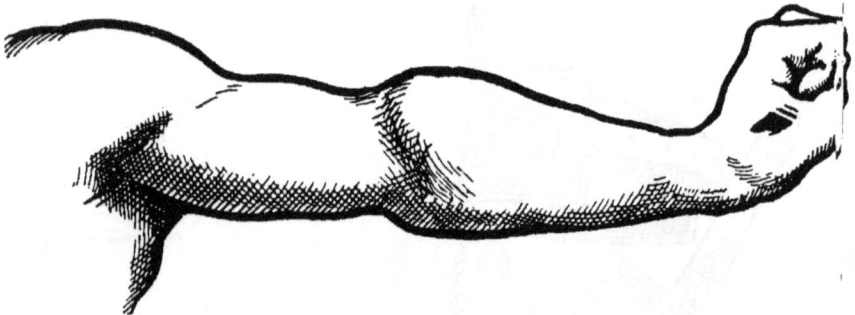

Fig. 202.—The most advantageous angle for the wrist to give a strong grip.

position, we promise improved power in the grasp
when the ankylosis has been attacked by the wrench
and the wrist placed in extension instead of flexion.
The forearm is then suspended in a sling or halter.
The management of abscesses is the same as at other
joints. Excisions and erasions are no more demanded

than elsewhere, and, as at the tarsus, are more mutilating in their results than at the larger joints. As a life-saving measure they may be demanded, as may amputation.

CARPAL, METACARPAL, AND PHALANGEAL DISEASE.

Chronic disease of bones of the hand, apart from wrist-joint disease, is very rare except in children and young adults. In most cases inherited syphilis is acknowledged or reasonably suspected. In the carpus and metacarpus the bones become thickened, and abscesses develop early. In the fingers, it is usually the proxymal phalangeal joint that is affected; the distal end of the first phalanx and the proxymal end of the second phalanx become thickened and the finger assumes a spindle shape, suppuration occurs and evidences of tuberculosis are found; nevertheless these cases respond to antisyphilitic remedies in a way that marks them as quite apart from ordinary tuberculous joints.

The mechanical treatment consists in splinting on the palmar surface. The operative treatment consists in subperiosteal erasion of the disease in the bones of the carpus and metacarpus, and in amputation of the fingers. If medicinal treatment is resorted to, rapidly-increasing doses of mercury and iodid of potash should be given until the constitutional effects are produced or the disease subsides.

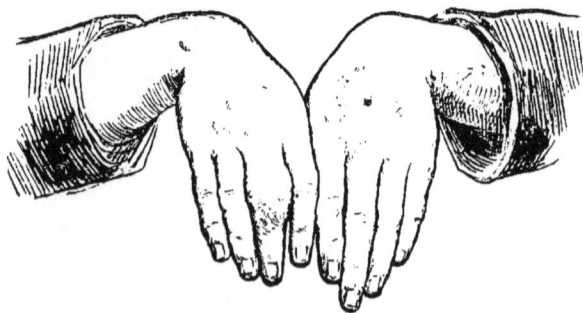

FIG. 203. -Disease of the metacarpal bone and of the finger-bones. characteristic of inherited syphilis.

RACHITIC DEFORMITIES.

The chief rachitic deformities with which the orthopedic surgeon has to deal are: Bowlegs, knockknees,
spinal curvature, and chicken-breast. The deformity
known as coxa vara will also be discussed, as well as
adolescent rickets. The rachitic condition which leads
up to these deformities will first be considered.

Rickets, or rachitis, is a disease of early childhood.
It is chiefly characterized by distortion of some or all
of the bones. The distortion is due to irregular
growth and deficient rigidity of the bony structure.

The etiology of rickets is obscure. Many causes have
been assigned by different observers, and any one, or
all, may be the chief etiologic factor in any given case.
The disease rarely develops after the third year of life;
by far the greater number of cases appearing during the
second year, during the months subsequent to the wean-
ing of the child. Evidence of rickets is occasionally
found at birth, and a certain number of rachitic deformities, especially of the long bones, develop at about
the period of adolescence.

A small number of rachitic children appear to owe
their disease to the last months of nursing when the
mothers have again become pregnant. Among the
wealthy we find rickets chiefly among bottle-fed babies,
or in those of syphilitic ancestry. Among the poor,
in addition to the above, improper food, bad air of
crowded and ill-ventilated rooms, lack of sunlight, and
uncleanly personal habits may be regarded important
causative factors in this disease.

In New York City fully $\frac{18}{20}$ of the rachitic children
observed by one of the authors (J. R.) were among the
poorer Italians, who live in dark, damp, and crowded

rooms; the children are seldom washed, and are permitted to eat anything that comes within their reach. Of the remaining $\frac{5}{20}$ fully $\frac{3}{20}$ were colored children of a class notoriously syphilitic; the remaining $\frac{2}{20}$ were about equally divided between bottle-fed babies and those of syphilitic ancestry.

In Chicago, where there are relatively fewer poor Italians, and where all classes of the poor are better fed and better housed, and enjoy more sunlight and pure air than in New York, there are comparatively few rickety children. The larger number of rickety children that have come under our observation are the children of colored people, presumably syphilitic; there are a considerable number from the children of Arabians and Syrians; and a relatively large number are children who have been nursed by pregnant mothers. Recently we have seen quite a large number that have been fed upon sterilized milk, and upon various brands of proprietary infant foods.

Parrot believed that all cases of rickets were due to inherited syphilis; that rickets was inherited syphilis; but the view of Fournier is probably nearer correct, that syphilis produces rickets indirectly by being one of the causes of malnutrition. In this connection it may be stated that many cases of rachitic deformity are due to irregular growth at the epiphyseal lines instead of a bending of the softened shaft of the bone, and that the seat of election of inherited bone-syphilis is at this point, the epiphyseal junction.

Macewen, of Glasgow, told Dr. Ridlon some years ago, at a time when he had operated upon over 1,200 rachitic deformities of the lower limbs, that diet appeared to him to play a much less important part than sunlight, pure air, and cleanliness; that the diet of the poor of Glasgow was practically the same as that of the poor in the Highlands, but that while having seen such an enormous number of rachitic deformities he had never seen one in a child reared in the Highlands.

It seems somewhat strange that rickets should be rarely found among the very poor in Ireland, but the observation is beyond question. Despite the fact that these children are underfed and illfed, housed in crowded and dirty hovels without floors or means of admitting sunlight or pure air, they rarely become rachitic. It has been observed that rickets occasionally is seen in the children of men who have been across to England for the harvest season and presumably have been exposed to venereal infection. The query naturally arises : Do the children of Ireland escape rickets because of the moral purity of the lives of their parents ?

In Japan, where moral purity is not counted as one of the crowning virtues, and where syphilis, possibly contracted from foreign sailors, is not unknown, medical missionaries find no rickets. They account for this on dietetic and hygienic theories. The Japanese live much of the time out of doors ; their houses are of so flimsy a build that ventilation is perfect ; they bathe every day, and from the earliest childhood fish forms an important part of the dietary. Mothers in childbed eat fish, and nursing children are fed upon fish as soon as they receive any food except the mother's milk. From this, one must conclude that syphilis does not always beget rickets, despite the fact that while the native African is free from rickets the Americanized and syphilized African is the father of rachitic children.

The infectious diseases of childhood, especially whooping-cough, and the digestive disturbances of infancy often precede the onset of rickets ; but perhaps the most that can be fairly said upon the subject of

etiology is, that anything which causes malnutrition
in a young child may lead up to the development of
rickets.

Even less is known of the pathology than of the

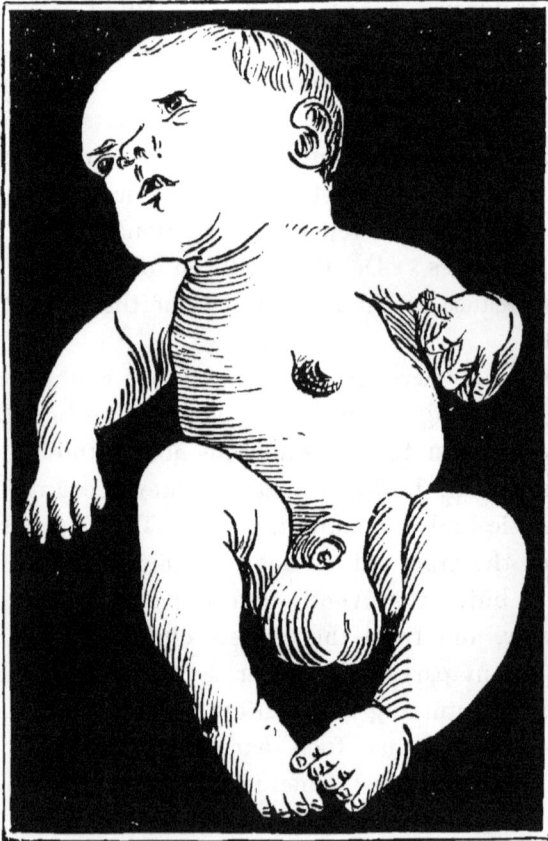

FIG. 204. Congenital rickets, showing bowlegs.

etiology. The new bone formed beneath the periosteum is soft and deficient in earthy matter, and in some
instances the animal matter does not yield gelatin on
boiling. The epiphyses are enlarged, and ossification

proceeds slowly and irregularly. The border of ossification, instead of being a clearly defined straight line,
is serrated, the new bone shooting into the cartilage
and the cartilage extending for some distance along
the shaft of the long bones. The medullary cavity is
advanced beyond the border of ossification and is filled
with a reddish, pulpy material. The lamellae of the
bone are loosely imposed, one upon the other, and in
some fresh specimens may be peeled off. The same
defective development is found in the flat bones.

The liver, spleen, and lymphatic glands are often
enlarged, and the quantity of white brain-substance is
increased. The muscles are pale and flabby and the
ligaments are relaxed.

The early symptoms of rickets· are indefinite, and the
onset insidious. The child is restless and yet not inclined to play about ; is peevish and yet is happier
when left alone than when held in the arms or carried
about ; tosses about in sleep, kicks off the bedclothes, and
perspires much about the head. The appetite is good,
but the child does not thrive ; there is some looseness
of the bowels, or constipation and diarrhea alternate ;
the abdomen becomes prominent ; the bone-ends, particularly at the wrists and ankles, become enlarged and
may be tender on moderately strong squeezing ; the
tibias may bow outward, or forward, or both outward
and forward ; the femora may bow in the same direction ; there may be knockknee from elongation of the
inner condyle of the femur, or outknee from elongation
of the outer condyle, or either deformity may result
from abnormality of the upper end of the tibia, or one
knee may be distorted inward and the other outward.
With bowlegs the feet are often flattened, and with
severe knockknee the knees are often held somewhat flexed ; with severe deformity in the legs and
thighs there is usually some flexion at the hips, the

pelvis is tilted forward, and there is lordosis of the lumbar spine. Associated with this tipping forward of the pelvis the weight of the body is brought to bear through the spine upon the sacrum in such a way that the sacrum is displaced forward and downward from the iliac bones, sometimes even to the full thickness of the bone, by that much diminishing the diameter of the pelvis. At times also the acetabulae approach each other, pushing the symphysis pubis forward and narrowing the pelvis laterally. The pelvic circle then becomes somewhat heartshaped, the point being to the front.

FIG. 205.—Congenital rickets, showing knockknees.

The sternum in the same way points prominently forward, the ribs sinking away on either side until the chest resembles the keel of a boat (pectus carinatum), or the breastbone of a fowl (chicken breast). At the junction of the ribs and their cartilages there is often so decided a thickening or beading that it can not only be felt, but can be seen. This is called the rachitic rosary. The lower border of the ribs flares out, because of the enlarged spleen and liver and the distended bowels. The spine is often bowed backward in the dorsolumbar region, so much so as to closely resemble the kyphosis of spondylitis, and if it is held rigid because of the tenderness of the bones the diagnosis between a spondy-litic and a rachitic curvature may be very difficult. The bones of the arms may be bowed like unto the legs, and the scapulae and collar-bones may also be dis-torted. The head is often of a characteristic appearance. It is flattened at the top and at the sides, the occiput

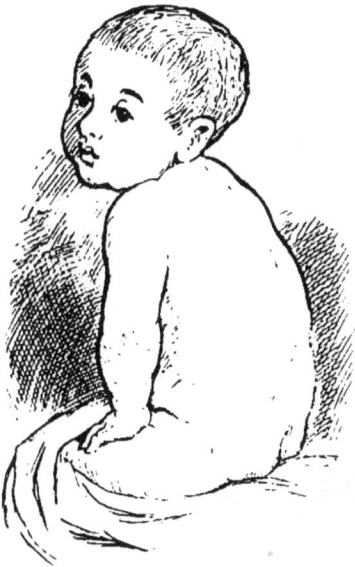

FIG. 206.—Rachitic curvature of the spine.　　FIG. 207.—Bowlegs in an infant.

and frontal bosses are prominent, and the head viewed from above suggests a square (caput quadratum). The face is small when compared with the head as a whole.

The lax ligaments, weak muscles, and tender bones render weight-bearing difficult, and these children consequently walk late, rarely before the fifteenth month, and often not before the end of the second or third year. Contrary to the general impression, children that walk at an early age do not develop bowlegs and knockknees, while those that walk late are very subject to these deformities. The teeth usually appear late and at irregular intervals; they are poorly formed and soon decay. Sooner or later solidification of the soft bones takes place, the process is usually a very rapid one, and the bone becomes unnaturally hard and ivory-like.

It is pretty generally believed that rachitic deformities

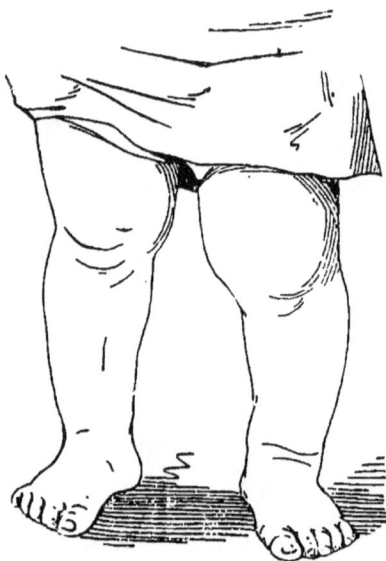

FIG. 208.—A case of bowlegs affecting the upper
third of the leg bones.

in the bones, particularly the outward bowing of the tibias, disappear. We are of the opinion that actual straightening of the bones does not take place, and that the true curve always remains, but there can be no doubt that the deformity, as a whole, appears to be less. This, we think, is due to the filling in of the concavity by the muscular development, to the greater length of the bones, to the tightening up of the ligaments, and to an increased muscular strength which enables the patient to carry himself in the best possible position for obscuring his deformity. These victims never attain their full stature.

The prognosis as to life is good, except in those young cases complicated with bronchitis and laryngis-

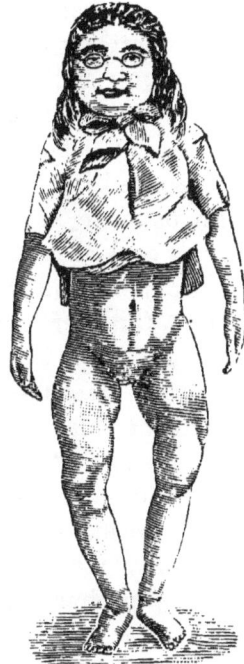

FIG. 209.—A case of unilateral bowleg.   FIG. 210.—A case of outknee in which there is little apparent deformity in the leg and thigh bones.

mus stridulus. The deformity, as a rule, steadily increases until the bones become solidified. In knock-knees the deformity may even increase after the bones have become solidified.

The general treatment is to improve the nutrition. Of chief importance are the following: Life in the open air and free exposure to sunlight. Light, dry, and well-ventilated rooms. Daily bathing, preferably in salt water or sea-water. An abundance of good food in which fish, fat, and fruit play a prominent part. The fish should be fresh and cooked with an abundance of butter, which, by the way, is the secret of the proper cooking of all fish. Fish should be given at least 3 times each week. Fat in the form of cream and butter in abundance wherever it can be used, and fried fat bacon. Butter will usually be taken more freely if un-salted or only half salted. Bacon should be cut thin, placed for a short time in iced water, carefully dried, and then placed in a hot pan without fat other than its own. When cooked it should be placed on grocers' ordinary brown paper to allow the excess of grease to drip off. Properly cooked it may be taken in the fingers without greasing them, and should be as brittle as a cracker or biscuit. Children will eat this with avidity twice a day.

Fruit in season should be given freely, and at such times as the child may crave it. Bread soaked in gravy or in milk should, as a rule, be avoided; and many of these children do not do well on oatmeal and other kinds of mush.

The only medication that has given us any satisfaction is phosphorus and codliver oil, with the iodid of iron in an occasional case and mercury and potassium iodid in the children of syphilitics. The various hypophosphites we believe to be useless. Phosphorus should be given in doses of from $\frac{1}{200}$ to $\frac{1}{100}$ grain. The elixir of free phosphorus is an acceptable preparation for private patients, and the oil of phosphorus in codliver oil is the most convenient in dispensary work. If the sirup of the iodid of iron is used it should be given with a free hand, and if mercury and potassium iodid are indicated they should be given in rapidly increasing doses until the physiologic effect is produced.

The mechanical treatment will be considered under the headings of the various deformities, as will the operative treatment.

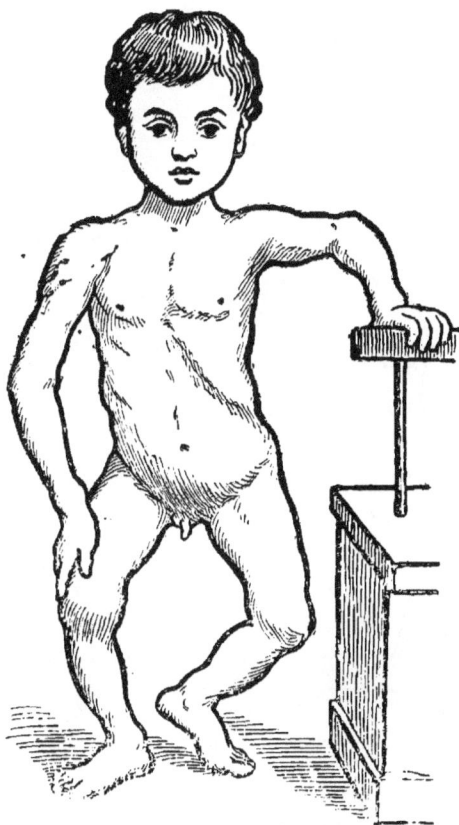

FIG. 211.—A case of bowlegs with inability to fully extend the knees.

Chicken-breast: Treatment by an elastic pressure-pad has been found of no avail, and little is ever accomplished in the way of actually correcting the chest-deformity by any form of treatment. A demand is often made that these cases be treated, and generally

enough gain can be had to satisfy the parents. We are
accustomed to have these children come to us daily for
exercises for a period of at least three months. The
exercises aim to increase the chest capacity, to reduce
the dorsal kyphosis (round shoulders), the lumbar
lordosis (sway-back), and the protruding abdomen.
They are as follows:

1. Back-lying, arms down, breathe deeply 10 to 20
times.

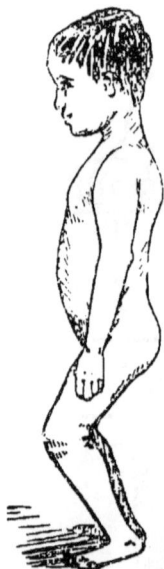

Fig. 212.—A case of bowlegs with
inability to fully extend the
knees.

Fig. 213.—Knockknees in an infant.

2. Back-lying, arms straight, grasp the ends of a stick
15 inches long, the surgeon pulls upward while the patient pulls downward, each alternately resisting, the
stick making excursions from the thighs of the patient
to the table above his head, 10 times.

3. Back-lying, arms stretched out and palms up,
breathe deeply 10 to 20 times.

4. Back-lying, grasp stick as in exercise 2, the patient
pulls upward while the surgeon pulls downward, each
alternately resisting, the stick making excursions from
the thighs of the patient to the table above his head, 10
times.

FIG. 214.—Knockknees and bowlegs in the same patient.

5. Back-lying, hands up, breathe deeply 10 to 20 times.
6. Back-lying, arms straight, the surgeon grasps the
patient's hands, each alternately pulling and resisting,
the arms make an excursion from the patient's thighs
horizontally to a point above his head, 10 times.

7. Back-lying, knee straight and stiff, foot extended, circle the legs, first one and then the other, 10 times.

8. Back-lying, knees held down, hands on hips, or stretched out, or locked at the back of the head, rise to the sitting posture, 10 times.

9. Back-lying, hands on hips, neck stiff and whole body rigid, the surgeon with his hands under the patient's head raises him 10 times.

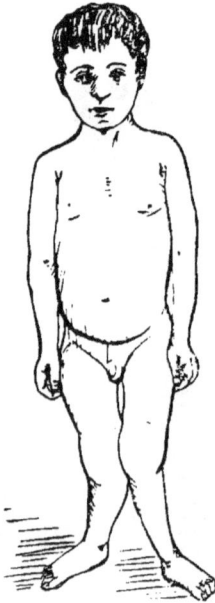

FIG. 215.—A case of knockknees.

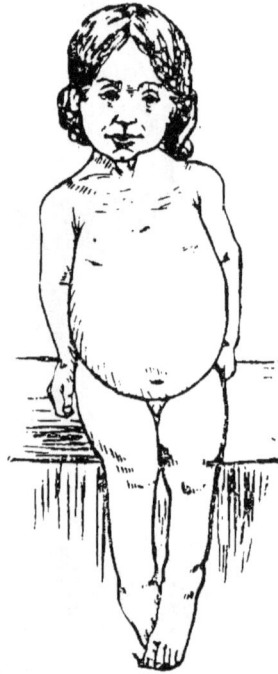

FIG 216.—A case of knockknees in which the deformity wholly disappears as soon as the knees are somewhat flexed.

10. Face-lying, hands on hips, or stretched out, or locked at the back of the head, legs held down, raise head and shoulders upward and backward, 10 times.

11. Face-lying, knees straight and feet extended, circle legs, first one and then the other, 10 times.

12. Horizontal-bar hanging, deep breathing, 10 to 20 times.

The order of the exercises, and their number and repetition must be graded to the strength of each case. All cases of rachitic deformity of the chest should be carefully examined for adenoid growths in the naso-

Fig. 217.—Rachitic deformity of the chest.

pharynx, or for other obstructions to the breathing; if found, all such should be removed. It is claimed by Redard, of Paris, that a careful examination will reveal some obstruction to breathing in all cases of this kind.

Rachitic Curvatures: Rachitic curvatures of the spine

may be posterior or lateral, or both posterior and lateral.

The lateral curvatures so closely resemble scoliosis that they cannot be distinguished from that condition. Indeed, most writers include rachitic lateral curvatures under scoliosis and class rickets among the causes of scoliosis. The treatment consists in holding the patient as straight as possible in a jacket or brace, as continuously and for as long a time as possible. The treatment is difficult of accomplishment and the results are far from satisfactory.

The posterior curvatures are usually in the dorso-lumbar region. The curvature is usually long and rounded, and if it is stiff, as it often is on account of the tenderness of the bones, it may be impossible to diagnosticate it from spondylitis unless there are other well-marked rachitic manifestations. The diagnosis, however, is not important, inasmuch as the mechanical treatment is by a brace of the same form and used in the same manner as in a like spondylitic curvature. The rachitic posterior curvatures always straighten, that is to say, the deformity is fully corrected.

Bowlegs: Bowlegs consist of a bowing outward or forward, or in both directions, of the bones of the leg, often accompanied by a bowing of the femur in the same direction, and associated with lax external lateral ligaments at the knee-joint. Occasionally one sees true outknee, due to an actual lengthening of the external femoral condyle or the outer side of the head of the tibia. When outknee is associated with an outward bowing of the femur and leg-bones, the limb, taken as a whole, may form an almost complete semicircle. The outward curve in the bones of the leg is usually near the upper or lower epiphyseal lines, although it may be in the middle of the shaft. The anterior curve almost always occupies the lower third or half of the tibia.

Mechanical treatment will correct the lateral deformity of the bones of the leg so long as the bone remains soft; but if the bones have hardened operative treatment must be resorted to. Anterior curvature of the tibia cannot be corrected by braces, and should be submitted to osteoclasis or osteotomy at once. Outknee can always be corrected by proper braces, even after hardening of the bones. If, however, the outknee be solely due to a lengthening of the outer condyle, it may be better, except in the very young, to do Ogston's operation, separating the outer condyle and allowing it to slip up far enough to correct the

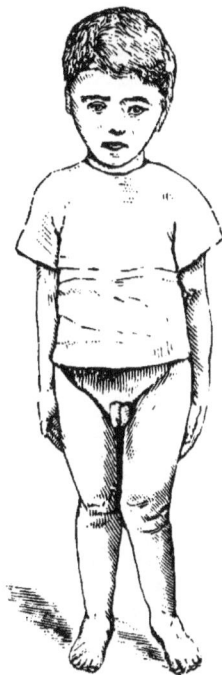

FIG. 218.—Unilateral knockknee.   FIG. 219.—The Thomas bowleg brace.

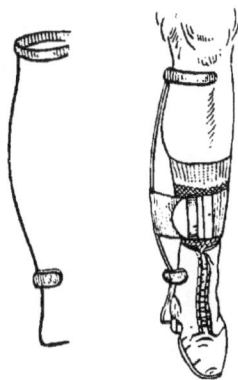

deformity. Bowing of the femur is seldom sufficiently severe to demand treatment.

In bowlegs, when the bones are very soft, it is often possible to correct the deformity by the use of plaster-of-paris. The plaster bandages should be rapidly applied, and before the plaster sets a piece of stiff board splinting is bound to the outer side of the leg (the con-

vexity of the curvature) by an ordinary muslin roller-
bandage. The roller and board-splint are removed
when the plaster has set.

The simplest and most efficient brace is a strip of flat
bar-iron reaching from the upper end of the tibia along
the concavity of the curve to the heel of the shoe where
it is bent to a right angle and passed into a hole or a
tube set in the heel of the shoe. The bent portion is
forged round so as to permit motion at the ankle-joint.
A padded iron band half encircles the ankle; this
may be completed by a leather strap. A like band is

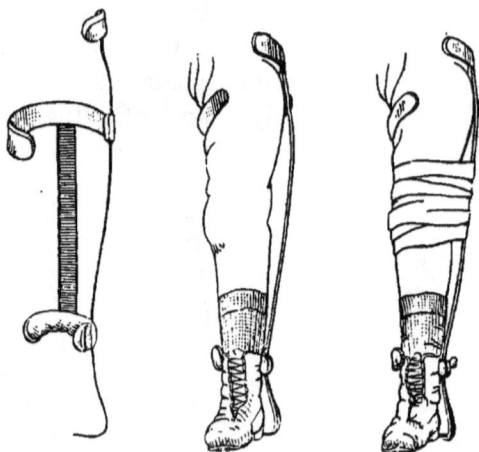

Fig. 220.—The Thomas knockknee brace.

placed at the garter-line below the knee, and a pad
rests against the head of the tibia. The curvature is
drawn toward the straight splint by a broad strap and
buckle or by an ordinary roller-bandage. When the
deformity involves the knee the side bar is extended
up against the thigh, and a posterior bar is added like
that in the knockknee brace to be described later.

After the deformity has been corrected for some
time, if any laxity of the ligaments of the knee remains
it may be well to put on the conventional leg-braces;

these consist of a strip of light steel on each side of the leg from the upper part of the thigh to the shank of the shoe, where it is riveted. Free anteroposterior joints are made at the knee and ankle, with pads at these points. A strong padded band joins the bars at the top around the back of the thigh, and this is completed across the front by a strap and buckle. A like band is placed at the garter-line. At times a strap encircles the leg and inner bar at the point of weakness in the tibia.

Knockknee: Knockknee, also known as inknee, results from a lengthening of the inner femoral condyle, or a shortening of the outer condyle, or a combination of both, although in some cases it is due to an elongation of the inner side of the upper end of the tibia. A certain degree of false knockknee may result from a laxity of the inner lateral ligaments of the joint. These cases of false knockknee can be cured by retaining the limb in the conventional leg-braces just described under bowlegs, until the slack in the elongated ligaments has been taken up by structural shortening.

Knockknee at any stage may be cured by proper braces, but brace-treatment on account of its long duration is hardly justifiable in cases past adolescence. In the younger cases it may be the best treatment in bilateral cases, but in unilateral cases the choice of treatment must depend upon the nature of the deformity. If the inner condyle be markedly lengthened, straightening by a brace results in lengthening of the entire limb, which may prove as great a disability as the knockknee. Such a condition should be subjected to Ogston's operation, to be described hereafter.

It is the custom of many surgeons and all instrument-makers who are called upon to treat cases of knockknee to apply the conventional leg-braces already described, with the addition of a large pad at the side of the inner condyle. This treatment for the correction of the deformity is absolutely useless. The secret of success in the brace-treatment of knockknee is in keeping the joint fully extended (straightened anteroposteriorly), while applying the lateral corrective pressure. The elongation of the inner condyle is only in the downward direction, not at all backward, and the full degree of deformity is apparent only when the joint is fully extended. As soon as flexion begins the corrective pressure-strain of the brace is relaxed and soon becomes nil; all knockknee deformity disappears long before

FIG. 221.—The Rizzoli osteoclast.

voluntary flexion reaches 90°. Therefore, when any degree of corrective pressure is applied by braces having a free anteroposterior joint at the knee the patient flexes his legs slightly and all pressure is removed. When, however, the deformity has been fully corrected by operative or mechanical means these braces serve well as a retentive measure.

The simplest brace for the treatment of knockknee consists of a strip of flat iron placed along the outer side of the limb, reaching from the greater trochanter to the sole, forged round at the bottom, bent at a right angle, and passed into a hole or a tube in the heel of the shoe; attached to this at the ankle is a padded iron band extending about two-thirds around the limb; at the upper part of the thigh is a like band, and these bands are joined by a strip of iron placed directly back of the middle line of the limb. At the top of the side bar is a pad from two to three inches in diameter. The limb is first bandaged to the posterior bar, securing the knee in full extension, and then to both bars, stretch-

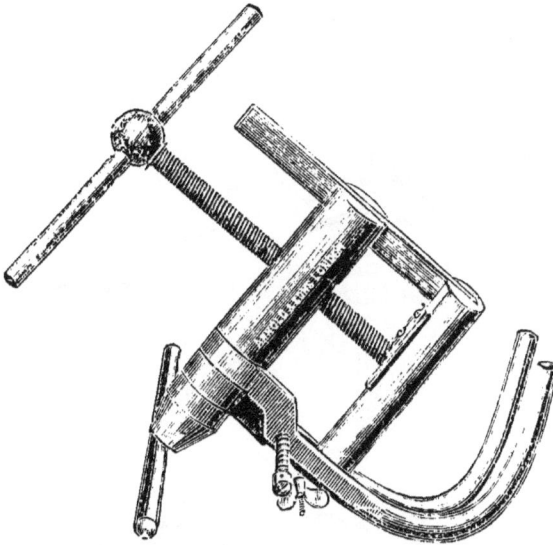

Fig. 222.—The Grattan osteoclast.

ing the knee in the direction of correcting the deformity. This corrected position is constantly maintained until the deformity is not only corrected but readily remains so; then the jointed retention-braces may be applied and joint action permitted.

Plaster-of-paris may be used in the treatment of knock-knee after the same manner as suggested in bowlegs, but it is less efficient. If it is used the limb must be firmly held by two persons while the plaster is setting,

270

or a strip of splinting may be bandaged on at the side
and another at the back of the limb.

The operative treatment of bowlegs and knockknees consists in an osteoclasis or an osteotomy of the bones at
the points of election. Osteoclasis is the making of a
simple fracture, either by hand or by the aid of a
specially devised instrument termed an osteoclast.
Osteotomy is the making of a compound fracture by
the use of cutting instruments for dividing the bone

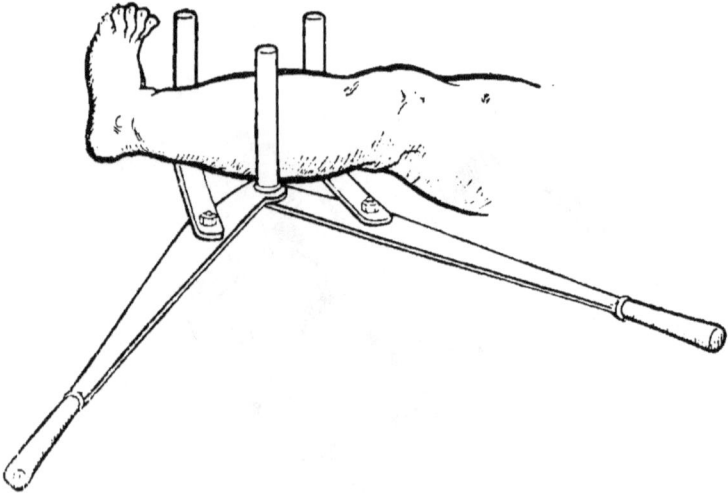

Fig. 223.—The Thomas osteoclast.

two thirds of the way, supplemented by manual fracture
of the remaining third of the bone.

Osteoclasis by the use of the hands of the surgeon
alone is not so easy a matter as at first appears. Comparatively few are able to break the leg bones of a child
of 2 or 3 years. When no other means are at hand the
leg can sometimes be broken over the edge of a table
by the surgeon throwing his weight upon it. In this
way it is not possible to fracture very accurately or very
near the end of the bone. The correction of knockknee

manually usually results either in tearing the outer lateral ligament or in partially tearing off the epiphysis. In either instance, however, the case usually does well provided the limb is supported for a sufficiently long time.

Mechanical osteoclasis is accomplished by a specially devised instrument called an osteoclast. The numerous forms of this instrument exert a breaking force in one of two ways, namely, by screw-pressure, or by

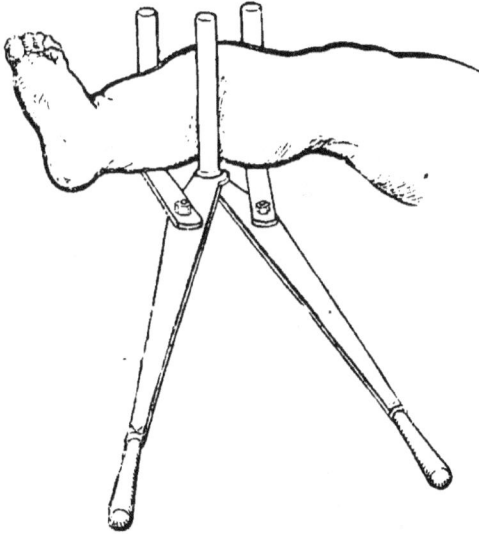

FIG. 224.—The Thomas osteoclast in use.

leverage pressure. Those working by screw-pressure are the most powerful and the most acurate; those working by leverage are the most rapid in their action, and consequently do less damage to the soft tissues covering the bone. .

The Rizzoli osteoclast has been extensively used because of its simplicity and cheapness. It consists of a heavy bar of steel, 1 inch by ⅞ inch, and 15 inches long, through the middle of which plays a screw with a handle at one end and a padded crutch at the other. On either side of this screw two steel loops are made to slide on the bar and fasten with thumbscrews. These loops are padded, and of sufficient size to admit the limb about to be broken. The instrument always breaks the bone transversely at the point where the crutch bears. The objection to the instrument is the manipulation to which the fracture is subjected in removing the instrument from the limb. In case the instrument is poorly constructed, and weak at the screw-hole, it may bend when in use, and its removal from the limb can not be effected except with the aid of a blacksmith.

Cabot's osteoclast is a modification of the Rizzoli by changing the loops into hooks, rendering its removal possible even if it should bend when in use.

Grattan's osteoclast is the one most generally in use among orthopedic surgeons in England and America at the present time. It consists of a strong post resting on a cross-bar. Upon the post are hinged two strong hooks, made to separate or approach each other by set-screws. Through the post plays a strong screw with a handle at one end and a pressure-bar at the other. This pressure-bar is steadied by a guide playing through a notch in the top of the main post. The instrument is not padded, but made of polished steel. The pressure-bar is ovoid on cross-section, with the smaller end of the oval directed away from the post and towards the limb to be broken. The instrument is the most powerful of any osteoclast that we have used.

The Thomas osteoclast consists of two steel bars joined in a hinge. At the point of the hinge an upright post is raised for the middle pressure-bar. A movable arm is attached to each bar a short distance back from the hinge; at the free ends of these arms on each a post is raised, and these form the outer pressure-bars. When the long bars are opened the outer pressure-bars go to the far side of the bone to be broken, while the middle pressure-bar impinges against the convexity of the curve; then when the long bars are made to approach each other the outer pressure-bars draw near while the middle pressure-bar goes out and the fracture is effected. The action of the instrument is very rapid, but it is not possible to break very close to the end of a bone.

The Ridlon osteoclast was devised with the idea of combining the accuracy of the Grattan with the rapidity

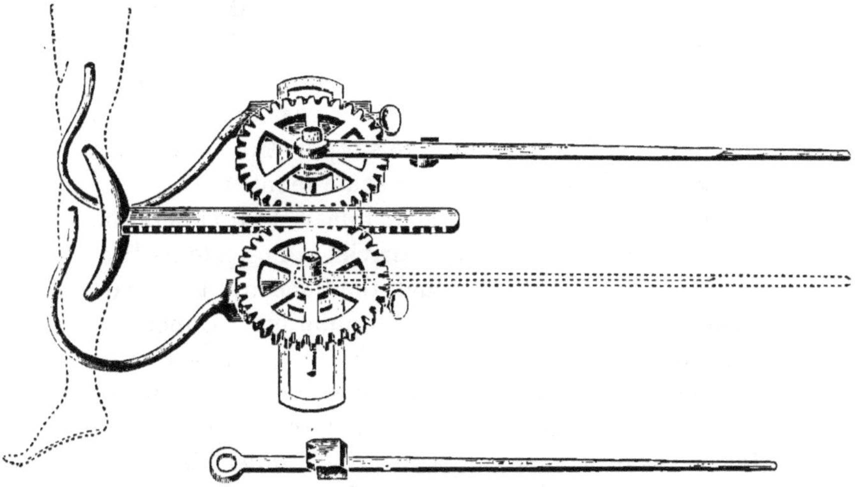

FIG. 225.—Ridlon's osteoclast.

of the Thomas. It consists of a flat steel bar, raised somewhat upon a rest; on the bar slide two hooks, fastened by thumbscrews; raised slightly above the bar are two toothed wheels; these are turned by two removable handbars; between the toothed wheels plays a toothed bar, at the end of which is the pressure-bar. The limb to be broken is laid in the hooks, and they are placed as desired and made fast; then the pressure-bar is run down against the limb, and the handle-bars adjusted at a convenient angle; when these are brought

together the pressure-bar is driven forwards by the turning of the wheels and the fracture is effected.

The Robin, the Colin, and the Lorenz osteoclasts are
complicated, expensive, and less efficient than those
already described. They are not used by English and
American surgeons.

After osteoclasis, the limb is wrapped with cotton-
wadding bandages and covered with a plaster-of-paris
dressing. After fracture of the bones of the leg the
plaster-dressing should extend from the toes to the
middle of the thigh. After fracture of the femur it
should extend from the toes to and around the waist.
Fixation of the fracture should be maintained from 4
to 5 weeks. After removal of the plaster in cases of
knockknee it is usually well to support the limbs with
the conventional leg-braces for some months.

Osteotomy is of two kinds: linear and cuneiform.
Linear osteotomy consists in driving an osteotome two-
thirds or three-fourths through the bone and then frac-
turing the remainder by manipulation. Cuneiform
osteotomy consists in removing a wedge-shaped piece
of bone from the convexity of the deformity by means
of a chisel and breaking the remainder by manipulation. The osteotome will be described under Macewen's
operation for knockknee. The chisel used in cuneiform
osteotomy is shaped like a carpenter's chisel, being a
flat piece of steel with parallel sides and at the end
beveled on one side by grinding to an edge.

Linear osteotomy may be the operation chosen for
the correction of the lateral deformity of bowlegs. The
section is made at the point of greatest deformity, and
after the manner to be described later as Macewen's
operation for knockknee.

Cuneiform osteotomy may be chosen for the correction of the anterior deformities. A wedge is removed
from the part of greatest convexity by means of the

chisel. The thickness of the base of the wedge depends upon the sharpness of the curve. The wound through the soft parts is in the line of the shaft of the bone; this is held open by retractors and the wedge removed transversely to the long axis of the bone. As a rule the flat side of the chisel is towards the longer portion of the bone and shavings are pared off from that side towards which the beveled side of the chisel looks. When a sufficient wedge has been removed the bone is broken by manipulation.

Osteotomy for knockknee, in so far as we know, was first performed by Ogston of Aberdeen, Scotland, on May 17, 1876; he sawed off the inner condyle of the femur. All operations on the inner condyle are modifications of this. On February 2, 1878, Macewen, of Glasgow, Scotland, first did the operation known by his name, using the os. teotome and making a section of the f e m u r above the condyles.

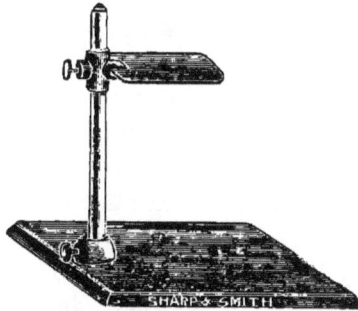

Fig. 226.—Ridlon's sacral table for supporting the hips when applying a plaster spica bandage.

Ogston's operation consists in the following procedure: The patient is fully anesthetized; the limb rendered bloodless; the leg is fully flexed on the thigh, and the thigh rotated somewhat outward; a long tenotomy knife is introduced flatly, two or three inches above the tip of the inner condyle and pushed downwards, forwards and outwards until the point is felt in the inter-condyloid space; the cutting edge of the knife is then turned directly towards the bone, and as it is slowly withdrawn all the soft tissues are divided down to the bone, and the external wound made sufficiently large to admit the blade of an Adams saw. This saw has a long narrow shank, a cutting edge about an inch and a half in length, and a blunt point. The saw is introduced along the canal made by the knife; its cutting edge directed towards the bone and the condyle sawed through about three-quarters of its thickness. The saw is then re-removed; the leg extended upon the thigh; and with a sudden forcible effort in the direction of straightening the knockknee the undivided portion of the condyle is fractured; the fragment slides up and the deformity is corrected.

The operation throughout must be strictly aseptic. The wound is closed without drainage; the dressings are applied; and the fractured bone is put up in a plaster-of-paris spica, or a wooden side-and-back splint.

The objections to the operation are, that the joint is opened; sawdust remains in the wound, and perhaps also in the joint; and an irregular joint surface or fault remains at the point of fracture, due to the slipping up-ward of the separated condyle. The advantages are that a normal length of limb results when the deformity has been solely due to an elongated inner condyle, a matter of no small moment in unilateral cases of knockknee.

Reeves' operation aims to avoid the sawdust by using an osteotome in place of a saw. The osteotome is driven along the same line as that followed by the saw in Ogston's operation, and the operator aims to stop short of the encrusting cartilage, and to leave the joint unopened. Most operators, however, believe that the joint is always opened in this operation.

Chiene's operation aims to avoid opening the joint by removing a wedge of bone from the inner condyle by means of a chisel. The base of the wedge includes some considerable portion of the inner part of the epiphyseal line, and its apex approaches the intercon-dyloid notch. After the removal of the wedge of bone the inner condyle is folded back against the shaft. The objections to this operation are, that it is difficult to perform; that it is difficult to estimate the necessary thickness of the base of the wedge; and that arrest of growth sometimes results from the injury to the epiphyseal line.

Macewen's operation divides the femur above the condyles by means of an osteotome. The Macewen osteotome is an instrument of the chisel order, beveled on both sides so as to resemble a slender wedge. The handle and blade form one piece. The handle is octagonal. At the top of the instrument is a rounded projecting head. One of the surfaces of the blade is marked in half inches. The whole instrument is finely burnished. The apex of the wedge ends in a cutting edge ground beveled to the sides of the blade like the cutting edge of a pocket-knife. Indeed, the blade of the osteotome closely resembles the cross-section of the blade of such a knife. A razor-edge does not have this second bevel, and the osteotome does not have the razor-edge, but the cutting-edge must be sufficiently sharp to pare the finger-nail.

The instrument is made from the finest Stubbs' steel, forged at a low heat, and tempered between a carpenter's tool and an iron-worker's tool. An inch of the tip is brought to a finer temper than the remainder of the blade. If there is any doubt about the temper it should be tested on a hard ox bone; the edge must neither turn nor chip. A set of osteotomes usually comprises three, the blades being of different thicknesses. The thicker instrument is always used first, followed by a thinner one when necessary. The osteotome is driven into the bone by a heavy mallet. Macewen prefers one of lignum vitae.

The patient must be profoundly anesthetized lest any movement of the muscles produce a larger wound than necessary. The limb is rendered bloodless by an Esmarch bandage. Macewen prefers to use the bandage as a constrictor instead of the elastic cord. The limb is placed upon a sand-pillow. The pillow is 12 by 18 inches, and only moderately filled with sand, so that the limb can be embedded in it. The pillow is moistened before the operation to prevent dust arising and to render the sand more cohesive. The point for the incision of the soft parts is on the inner side of the thigh, half an inch in front of the adductor magnus and on a level with a line drawn transversely across the thigh a finger's breadth above the upper border of the external condyle. The incision in the soft parts must be made parallel with the long axis of the thigh, and ought to be a sharp, clean, single incision, produced by one stroke of a good-sized scalpel. It is not desired that the knife divide the periosteum. Before the knife is removed the osteotome is introduced. After the removal of the knife the cutting-edge of the osteotome is turned transversely across the bone. The handle of the osteotome is grasped by the left hand of the operator, his thumb underneath the head of the instrument and the ulnar surface of his hand, or of his wrist, resting on the limb. The cutting-edge is brought in contact with the posterior border of the inner surface of the bone and directed towards the outer and anterior border, and in this direction driven through the inner hard portion of the bone, the soft portion, and up to the outer hard portion. After each blow of the mallet the osteotome is lifted slightly by the hand which grasps it so that it does not become firmly wedged into the bone. This lifting is not done by swaying the instrument from side to side, but by grasping it close to the limb, and swelling the muscles on the ulnar side of the hand against the limb. If it cannot be made to penetrate sufficiently far without wedging it should be replaced by a thinner tool. When the bone has been penetrated sufficiently far in the direction indicated, a second entrance is made from the anterior border of the inner side of the bone, with the cutting-edge directed outwards and backwards. In this manner the popliteal artery behind and the prolongation of the synovial sac in front are not endangered. When sufficient bone has been cut the remainder is broken by bending the limb in the direction of correcting the knockknee. In young subjects the limb is bent slowly with the intention of producing a greenstick fracture; in older and harder bones it is snapped by a sudden blow-like bend.

The wound is dressed without drainage. Macewen closes it with a small piece of carbolized Lister protective; others use carbolized rubber tissue; and others seal the wound with collodion and cotton or gauze, after taking two or three superficial stitches. In either case an aseptic dressing is placed around the limb for a considerable distance above and below, and the fracture immobilized in a plaster-of-paris spica or a back-and-side board splint.

After the patient has recovered from the anesthetic the surgeon must assure himself that the circulation, the sensation, and the motion in the toes are normal. At the end of a fortnight it will usually be found that the wound has healed, and the antiseptic dressings can be removed. Immobilization of the fracture must be continued, however, until union is perfect.

Macewen believes that the advantages of the operation are that no bone is removed; that both sides of the bone contribute to rectify the deformity, one being compressed while the other stretches, or gaps with the periosteum preserved over the gap. The operation is much more frequently chosen by surgeons generally than any other cutting operation and in bilateral cases is undoubtedly the best.

Adolescent Rickets: Adolescent rickets has generally
been considered as identical in its characteristics with
infantile rickets. Our observation is not in accord with
these views; nor with the opinion that it is found only
during the period of adolescence The so-called rachitic
deformities of adolescence that we have observed are
bowlegs, knockknees, and coxa vara. In the adolescent bowlegs that we have seen the deformity is confined to the proximal end of the leg bones and is usually unilateral; whereas infantile rickets shows the
deformity in any part of the leg bones in the following
order of frequency : in the lower third; in the lower
third and the upper third about equally; in the upper
third alone; in the shaft; and the deformity is usually
bilateral. The adolescent bowleg is then more strictly
an outknee, the opposite of inknee (knockknee), than a
true bowleg. The difference in the bony deformity
between the adolescent and the infantile forms may be
seen in the Röntgen pictures of these two deformities.
Adolescent knockknees do not show the peculiarities
equally well. The deformity seems to begin in the
upper end of the tibia, but as soon as any deformity
has developed the line of gravity falls too far to the
outer side of the joint, the inner lateral ligaments
stretch, the line of gravity goes farther out, weight is
carried on the outer femoral condyle and its growth is
retarded while the inner condyle relieved from weight-
bearing grows abnormally long. Similar changes take
place in the articular surface of the tibia. Coxa vara
consists of a change in the relation of the head of the
femur to the shaft of that bone. The head gradually
sinks to or below the level of the greater trochanter
and nearer to the shaft of the bone. In some cases the
femoral neck seems to bend, while in others it seems
to actually disappear to a very considerable extent and
the globular head may appear to lie against the inner

FIG. 227.—Coxa vara, showing depression of the head of the femur and shortening of the neck. This is the first case described under adolescent rickets.

surface of the upper end of the shaft of the bone. At
times the trochanteric region appears thickened. The
process is usually a gradual one and is unaccompanied
with pain or any special sensitiveness.

The main characteristics of this so-called adolescent
rickets are: The condition appears somewhat later
than the usual period of infantile rickets, and in chil-

FIG. 228.—Showing range of flexion in the boy with coxa vara. The first case
here reported.

dren who have had a healthy infancy; it is more often
than otherwise unilateral and confined to the proximal
ends of the long bones; it is not accompanied by the
usual symptoms and signs of infantile rickets, such as
enlarged wrists and ankles, enlarged abdomen, deformed
chest, square head, etc.; it progresses for several
months, or even for 2 or 3 years, when the progress

ceases and the deformities thenceforth remain unchanged. Correction of the deformity after the progress has ceased gives permanent results; correction
before this time is usually followed by a return of the
deformity. A traumatism occurring during the progress of the deformity may render the bone-end sensitive for some weeks and make a differential diagnosis
from tubercular joint-disease somewhat difficult. It is
only by a careful consideration of the history of the

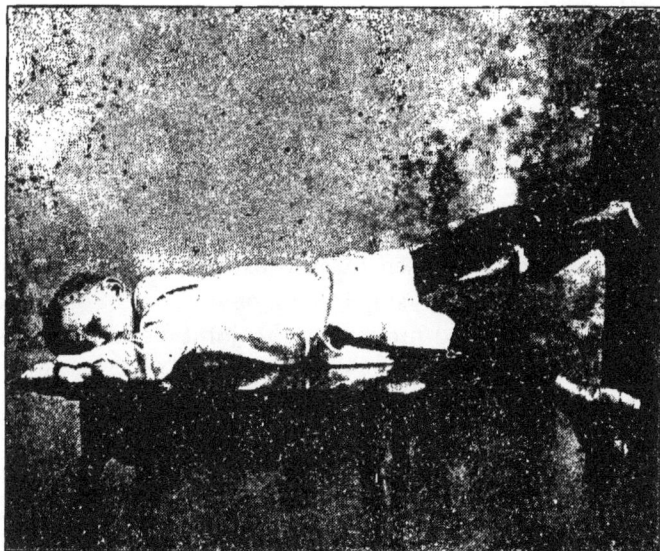

FIG. 229.—Showing the range of extension in the boy with coxa vara. The first
case here reported.

case and the essential diagnostic symptoms of tubercular disease that one may be reasonably sure. The Röntgen picture may help very materially, but observation
for some weeks may be required to render a conclusive
opinion. The diagnosis of coxa vara is of course more
difficult than that of bowlegs and knockknees. The
following cases will illustrate that difficulty:

A boy, 5 years old, the second of three healthy

children of healthy parents, was noticed limping slightly in January, 1896. He had always been well and was an unusually active and robust child. There was no complaint of pain and no disability except the slight limp. Examination showed $\frac{1}{8}$ inch shortening of the limb measured from the anterior superior spine of the ilium to the inner malleolus; no shortening measured from the tip of the greater trochanter to the outer malleolus; and upward displacement of the greater trochanter equal to the amount of the shortening. There was no restriction to motion in any direction. From this time on we measured the limb once or twice each month for 6 months and the $\frac{1}{8}$ inch shortening gradually increased to $\frac{3}{4}$ inch. There were no other symptoms. The diagnosis was coxa vara; and this was confirmed by Dr. L. L. McArthur, of Chicago.

In June, 1896, the patient went to Yonkers, New York, where he amused himself by repeatedly jumping down three or four steps to a stone walk. Soon the limp increased and motion at the joint became somewhat restricted. He was examined by Drs. Newton M. Shaffer, Virgil P. Gibney, and Royal Whitman, of New York.

Dr. Shaffer regarded the case as one of chronic disease of the head, neck, and acetabulum.

Dr. Gibney believed it to be a bending of the neck of the femur.

Dr. Whitman was inclined to believe it was a tuberculosis of the neck of the femur.

The boy was at once returned to our care. We found much the same symptoms as were recorded by Drs. Gibney and Whitman. Our previous observation of the case, however, led us to doubt the conclusions of the New York surgeons in so far as the question of tuberculosis was concerned and to hold to the previous diagnosis of coxa vara. Nevertheless, as a precaution against error on our part a long traction hip-splint of the pattern approved by Dr. Shaffer was applied, this also acting as a perineal crutch and protection to the joint as advised by Drs. Gibney and Whitman. Within 6 weeks all restriction to the normal range of motion had disappeared and the boy was in precisely the same condition as before going to New York. From this time on there was no change except that the muscles of the leg were less well-developed than those of the other leg.

About 7 months later Drs. Shaffer and Whitman again examined the patient.

Dr. Shaffer was confirmed in the opinion that the boy had a tuberculous hip.

Dr. Whitman was in doubt as to the pathologic process that caused the bending of the femoral neck.

The condition of affairs remained unchanged. In May, 1897, after having worn for eleven months the long traction hip-splint, commonly employed to immobilize and protect in tuberculous arthritis of the hip-joint, photographs were made to demonstrate the range of motion at the joint in flexion and extension. Fig. 228 shows the boy easily able to flex the limb, held straight at the knee, so that the foot approaches the ear, Fig. 229 shows equally full hyperextension. One would hardly expect to find so free a range of motion in a healthy joint after having been constantly confined in a long traction hip-splint for eleven months, and it positively excludes the presence of tuberculosis in or near the joint, or its existence in those parts at any previous period. The condition of affairs remaining unchanged at the end of two years, the shortening being $\frac{3}{4}$ inch, the adhesive plasters used in making traction were dispensed with, but the splint was continued —used only as a perineal crutch. Three years from the beginning of the observation, in January, 1899, the

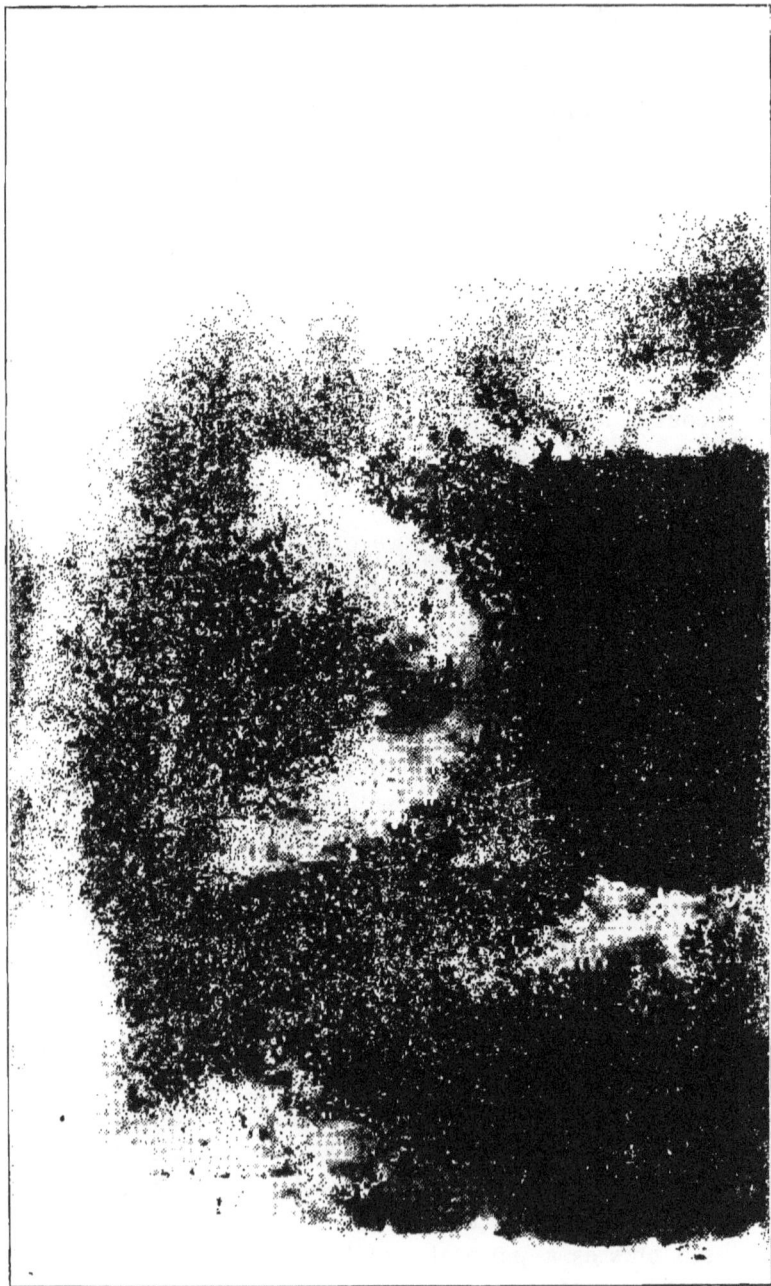

FIG. 230.—Adolescent rickets, showing depression of head of right femur (left side of illustration), absorption of neck, and thickening of trochanteric region with lessened density. This is the second case described in the text.

FIG. 231.—Adolescent rickets, showing changes in the upper end and of the tibia. This is the second case described in the text.

perineal crutch was put aside. When last examined, in May, 1899, the condition was unchanged.

E H , male, 12 years old, was first seen by us on February 1, 1897. The family history is free from tuberculosis and rickets. The patient never had any special or serious sickness. When about 7 years old it was noticed that he limped and stepped on the toe of his right foot. About a year and a half later, when 8½ years old, he was examined by Dr. Nicholas Senn, of Chicago The mother, an intelligent woman, thinks there was some question as to the diagnosis, but she was told that it was incipient hip-disease. She is sure that there was no difference in the measurements of the two legs at that time. She was instructed as to the treatment, which was weight-and-pulley traction at night, but was told to let the boy continue out of-door exercise during the days. No brace was applied. Later on the boy was circumcised.

Six months ago the mother first noticed outward bowing at the right knee, and that he leaned to the right in walking. There had never been any pain or tenderness, or night cries. The patient stands with the right knee slightly flexed and bowed outward, and the right hip somewhat lower than the left. The thigh can be flexed to a right angle; rotation is very normal; adduction normal, abduction more than half the normal extent; hyperextension normal. There was no muscular spasm, and the restriction to motion appeared to be bony. Measured, anterior superior spine of ilium to inner malleolus, 1 inch shortening; from tip of greater trochanter to external condyle of femur, no shortening; from tip of internal condyle to inner malleolus, ½ inch shortening. The circumference of the right thigh was ¾ inch less than the left, and the right ca'f ¼ inch less than the left.

The Röntgen photograph, Fig. 230, shows the condition of affairs at the hip. The head of the femur is depressed; the neck shortened; and the trochanteric region thickened and somewhat lessened in density. There does not appear any evidence of present or past disease of the hip-joint itself. Fig. 231 shows the condition of affairs at the knee. The upper end of the tibia is thickened; less dense; has irregular notches and projections at the sides, and three transverse lines somewhat resembling the single lines that appear at the epiphyseal junctions of bones in young children.

A comparison of these pictures with the Röntgen pictures of ordinary infantile rickets demonstrates the marked difference that exists between the two conditions.

FIG. 232.—Adolescent rickets, showing knockknee. The patient holds a Thomas knockknee brace in his left hand.

When much thickening takes place in the bone ends, whether at the hip or knee, but more especially at the hip, the range of motion will be diminished to some extent. At the hip, abduction is restricted more than the other motions, and next to abduction rotation is restricted.

Treatment during the progress of the disease may, and probably does, lessen the amount of deformity. At the knee, braces should be applied in the same way as

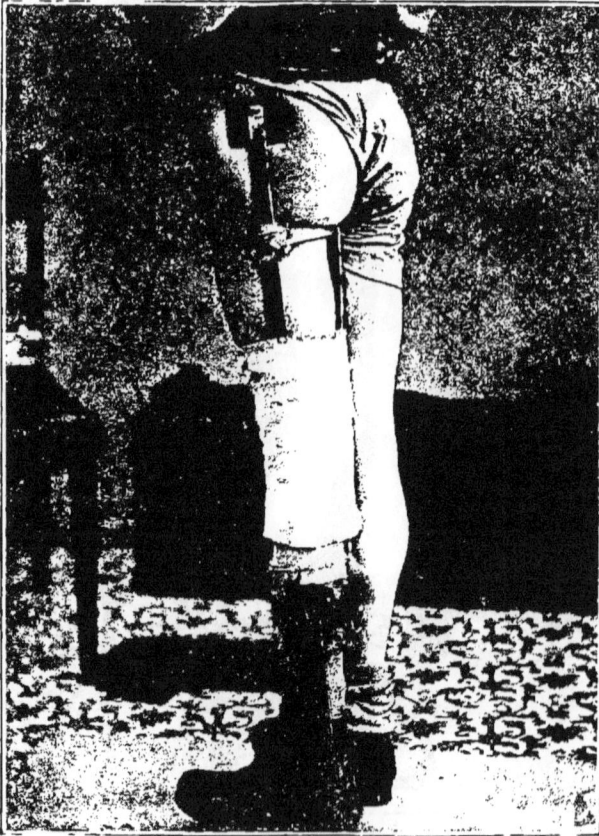

FIG. 233.—Adolescent rickets, showing the Thomas knockknee brace applied.

in the treatment of the same deformity in infantile rickets; at the hip, support should be given by some form of perineal crutch to relieve the femoral neck, weakened by disease and at a disadvantage from its right-angled position, from weight-bearing. When the disease has ended, the deformity at the knee can be corrected by osteoclasis or osteotomy as infantile rickets. The true shortening from coxa vara may be only partially corrected by osteotomy in the trochanteric region, but the false shortening from adduction and flexion can be fully corrected. The operation is precisely the same as that performed for the correction of adduction deformity resulting from hip-disease.

Clubfoot, talipes, is a deformity of the foot, consisting mainly of a distortion of the bones of the tarsus and of the foot as a whole in its relation to the leg. The simple varieties of talipes are equinus and calcaneus, varus and valgus; the compound varieties are equinovarus, equinovalgus, calcaneovarus and calcaneovalgus. These deformities may be either congenital or acquired, and when acquired are usually the result of infantile paralysis. Another deformity of the foot due to paralysis is cavus, or pes cavus as it is more often called. Still another acquired form is planus, or pes planus, due to inherent weakness in the foot, or to excessive or long-continued weight-bearing. The paralytic deformities, and those due to weakness, will be considered under another heading.

In talipes equinus the foot is plantar flexed, *i. e.*, extended on the leg, without lateral deformity, and can not be dorsal flexed. Calcaneus is the opposite deformity, the foot being dorsal flexed and the heel presenting, the patient being unable to extend the foot on the leg. In varus the foot is shortened in all its structures on its inner surface, and is both inverted and rotated inward from its normal relations to the leg. In valgus the foot is everted and rotated outward in its relations to the leg. The compound varieties present combinations of the above-mentioned deformities.

Nearly all the cases of congenital clubfoot present the compound variety—equinovarus. The foot is extended and the heel is drawn up; the foot as a whole is rotated inward and inverted, but the front portion of the foot consisting of the metatarsus and phalanges is bent further inward and is more inverted than the

tarsal portion of the foot. In severe cases the sole is
shifted from the horizontal to the vertical plane and
looks inwards and backwards, or directly backwards.
A few cases present the equinovalgus deformity.
Simple calcaneus is exceedingly rare; and simple
equinus is still more rare. We have not seen simple
varus (that is, varus without equinus) nor simple valgus. In the writings of the older surgeons frequent
mention is made of "talipes varus," but it is the vari-

FIG. 234.—Unilateral congenital
equinovarus.

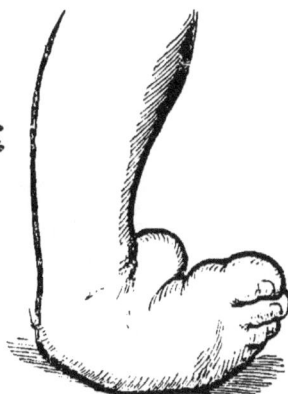

FIG. 235.—Congenital equinovarus
of marked degree.

ety now known as equinovarus and not simple varus
that is meant.

Congenital clubfoot is not of frequent occurrence.
Lannelongue found only 8 children born with clubfoot
in over 15,000 births, or about one in 1,900.

The etiology of clubfoot has been much befogged
by numerous and conflicting theories. At least 5 have
had the backing of the best men in the profession.
These theories are:

1. The paralytic theory, which received the support

of William J. Little, the father of orthopedic surgery in England. Because of the similarity between the congenital and the acquired forms it was assumed that congenital deformities were due to the same nerve-lesions that produced acquired clubfoot. Microscopic examination, however, does not reveal the changes in the brain and cord in cases of congenital clubfoot that

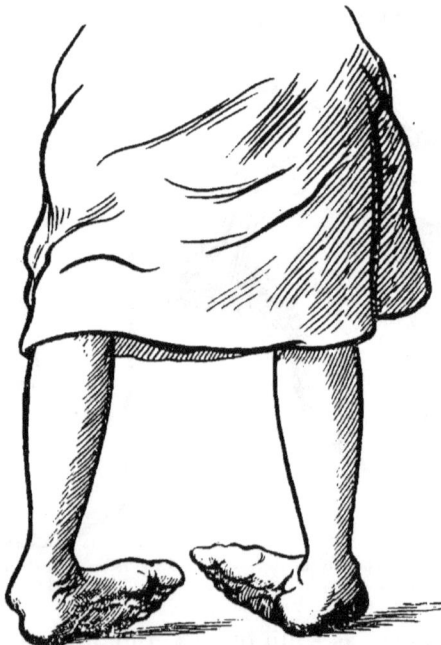

FIG. 236.—Bilateral congenital equinovarus. Deformity increased from walking in the deformed position.

are demonstrable when distorted feet are due to infantile cerebral and infantile spinal paralysis. The electrical reactions are not changed from the normal, and the voluntary motion, the color and temperature of the skin and the muscles are quite different from what is found in paralytic clubfoot.

Further, paralytic clubfoot is the result of years of

malposition, no case becoming really deformed in so
short a period as nine months. In offering the above
objections to the paralytic theory we do not mean to
say that a case of paralytic clubfoot which has been
present from birth may not have antedated birth. We

Fig. 237.—Congenital equinovarus before treatment.

believe that we have seen one such case: A girl of 9 years,
the only child of educated and observing people. The
deformity was described as a typical double equinovarus, and was present at birth. It had been treated

much of the time for the nine years by braces, manipulation and massage. No operation had been performed, and the parents had been warned against an operation by the late Dr. Roth, of London, in whose care the child was for some time. When we saw the case, the appearance of the feet was much as one would expect in a severe congenital case treated as this had been. The varus had been almost wholly corrected, but the front of the foot was abnormally broad; the foot, however, could be inverted and the head of the astragalus made prominent. A marked degree of equinus was present, indeed the feet could not be extended (plantar flexed) beyond the point where they were habitually held, and dorsal flexion could be demonstrated only to a very slight extent and when considerable force was used. The tip of the os calcis, for the attachment of the tendo-Achillis, was markedly prominent and displaced far inward; the skin and local temperature appeared to be normal; the electric reactions were not tested. It was assumed that it was a congenital case of the usual type and that the anterior leg muscles would rapidly regain their normal contractility and strength when relieved of strain and relaxed. To this end the tendo-Achillis was cut on both legs and the posterior ligaments of the ankle-joints ruptured so that the dorsal flexion of the feet was perfect. Now at the end of 5½ years, although the feet have been retained in the normal position, there has been no return of voluntary muscular control and no gain in voluntary dorsal flexion.

2. The second theory is that of abnormal (excessive) uterine pressure. This theory has been supported by most of the older writers and such prominent modern surgeons as Volkmann, Kocher, and Vogt, in Germany, and R. W. Parker, in England. The objections to this theory are that clubfoot is often found in instances

when there is known to have been an abundance ot
amniotic fluid, and absent when the amniotic fluid was
scant; in fact, that it bears no relation to the quantity
of fluid; that twins have been born, one with, and one
without, clubfoot; that there are hardly any other con-

FIG. 238.—Same case after treatment.

genital deformities that could be accounted for by excessive uterine pressure; and that the frequency of the
equinovarus deformity and the infrequency of other
clubfoot deformities is not accounted for by the theory.

3. The third theory is the opposite of the last, namely, it is the theory of the lack of normal uterine pressure. This was advanced by Luecke, who called attention to the fact that very many fetuses at some early period of their development present a greater or less degree of inversion of the feet and that this diminishes as full term approaches. He held that the lack of a foothold, so to speak, against which to kick in those cases where an excessive amount of amniotic fluid was present resulted in the feet remaining in the equinovarus position.

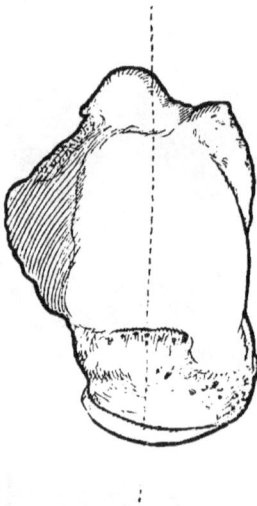

FIG. 239 —Normal astragalus. Showing relation of axis of neck to the tibial articular surface.

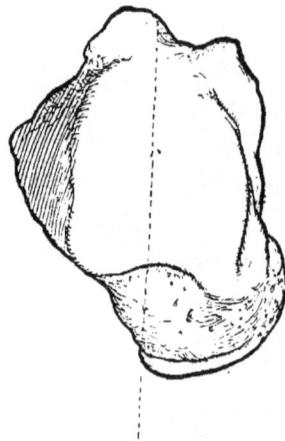

FIG. 240.—Deformed astragalus from a case of congenital equinovarus. Showing distortion of neck in relation to tibial articular surface.

This theory was developed and carried to its logical sequences by H. W. Berg, of New York. He says : "As soon as the joints are formed we find the thigh rotated out as far as possible, and flexed upon the body. The leg is flexed upon the thigh, but not completely, for this is prevented by the extreme rotation of the thigh which brings the inner border of the leg pressed against the abdomen of the fetus, the legs crossing each other a little below the middle. All the intrauterine pressure, therefore, is brought to bear directly upon the outer border of the leg and thigh, and also upon the dorsum of the foot. The result of this is that the foot is rotated in and extended (equinovarus) until the sole

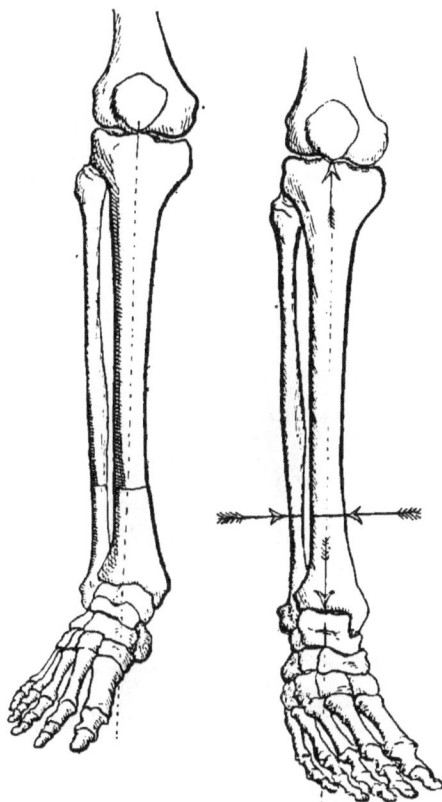

FIG. 241.—The drawing on the right shows the inward distortion of the bones of the tarsus in congenital equinovarus.

is almost on a line with the inner border of the leg, and lies against the body of the fetus, while the dorsal surface of the foot is on a convex curved line with the outer border of the leg, to adapt itself to the concave wall of the uterus. This, I believe, is a stage in the

normal development of every healthy fetus; and were
the extremities to remain in this position, all children
would be born clubfooted. But nature provided against
this by the outward rotation of the extremity, which
gradually takes place, carrying the leg away from the
position against the abdomen of the fetus; and when
this rotation is completed we find the extensor surface
of the thigh flexed and in relation to the body of the

FIG. 242.—Unilateral congenital equinovarus after treatment. Some varus
deformity still remaining.

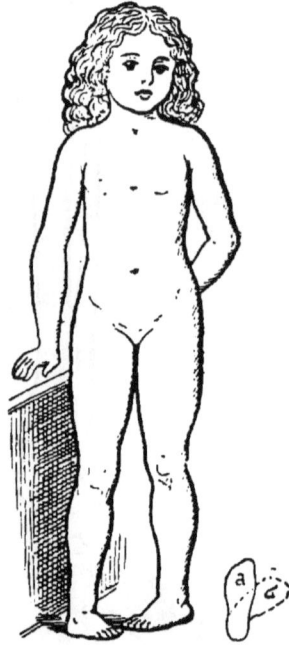

FIG. 243.—Intoe deformity remaining
after correction of the antero-posterior (equinus) deformity and
the lateral (varus) deformity.

child, while the legs are flexed upon the thighs, the
inner or tibial borders facing each other. Now the
soles of the feet lie against the uterine walls, and the intrauterine pressure is exerted directly upon them. This
produced extreme flexion of the foot upon the leg, together with an outward rotation of the foot; this move-

ment, from the constitution of the ankle-joint, accompanying extreme flexion. Thus is antagonized the varus or equinovarus existing hitherto. It is evident, then, that upon the completeness of the internal rotation or torsion which takes place in the lower extremity, depends the rectification of the early varus of the foot. Sould this rotation not take place at all, or be incomplete, the foot will continue to maintain its early relation to the body of the fetus and uterine walls, and the child will be borne more or less clubfooted. If this is so, we should expect to find in clubfooted children that the extremities are rotated outward. And this we do find upon examination. In all of the cases of congenital clubfoot (equinovarus) which I had seen since my attention has been directed to this subject, I have found that the thigh and leg, as a whole, were rotated out, and the tibia bent at its lower part, so that the feet are approximated to each other in addition to being in the clubbed position. All this is seen to be the result of nonrotation of the leg."

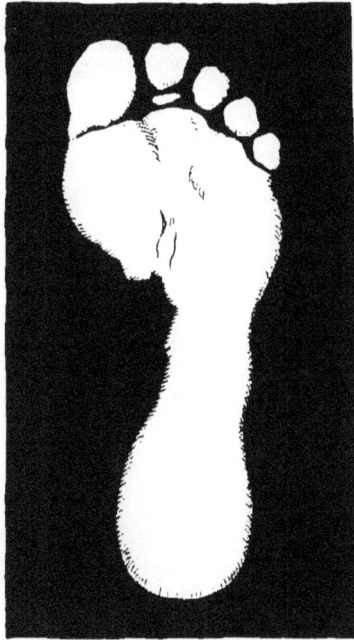

Fig. 244.—Imprint of a normal foot.

The chief fault in Berg's theory is that it only accounts for equinovarus; and does not account for congenital calcaneovalgus and congenital clubbed hands which we have seen in a patient suffering from congenital equinovarus. It is probable that at the time Berg wrote his paper he had never seen any other form of congenital clubfoot.

4. The fourth theory, advanced by Professor Eschricht, of Copenhagen, in 1851, is that of arrest of development. It is based on the observation that clubfoot deformities, like such other congenital deformities, as clubhands, webbed, supernumerary, and deficient

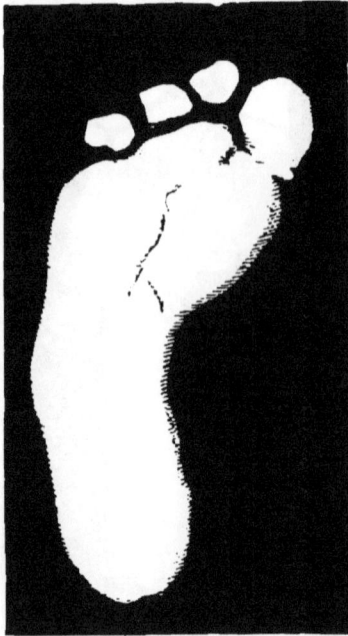

Fig. 141.—Imprint of a foot resulting in resolution from ... in the attending circular surface of the toe resulting after correction of the equinovarus deformity.

fingers and toes, constriction bands, amputations and deficient members and parts of members, harelip, cleft-palate, spina bifida, acephalus and other monstrosities, more frequently are found in males than females, more frequently in children born out of wedlock and in those whose mothers have had unusual worry and anxiety than otherwise; and more frequently in first children than those born at a later period. In support of the above, innumerable instances have been observed and recorded; an equal number have been observed controverting the theory. The following illustrates the point. Some years ago in New York a widower, the father of 3 normal children, met and married a widow, the mother of 2 normal children. There was no hereditary deformity in either family and these two, man and wife, were in no way related by consanguinity, but the result of the union was 3 children, all with double talipes equinovarus. This instance would appear to controvert each of the foregoing theories.

5. The fifth theory is that of heredity, in support of which the account of a French shoemaker, to be found in Valentine Mott's edition of Velpeau's "Surgery," translated by Townsend, may be related. It seems that the worthy shoemaker had double talipes varus, and his first four children all had double talipes varus, but the fifth child was born with normal feet. The shoemaker accused his wife. Arguments against heredity do not need to be advanced, for cases like the above are too infrequent to demand an extended argument.

In considering all of these theories the only conclusions that can be reached are that an individual case of clubfoot may be accounted for by any one, or by several, or by none of these theories, and that in most instances we do not know what causes the deformity.

As to the anatomy, we have seen neither dissections nor descriptions of the anatomy of congenital equinus or valgus.

R. W. Parker reports two cases of congenital calcaneus in which the tendons of the tibialis anticus and extensor proprius hallucis and the anterior ligament of the ankle had to be divided before the foot could be fully extended. In these cases the only change in the bones was in the upper facet of the astragalus, which was displaced somewhat forward.

In acquired deformities of the feet the change consists only in those alterations of the articular surfaces that arise from the malpositions, and such bony changes as are due to growth in the malpositions. We may then limit our consideration to the changes occurring in congenital equinovarus. These cases differ somewhat, inasmuch as the distortion differs in degree in different cases, and description of writers vary somewhat depending upon the case, or cases, dissected by them.

To Dr. Frank Hartley, of New York, we are indebted for the following description of a congenital case of 22 years' standing, the position being that of extreme equinovarus:

There was a large multilocular bursa covering the bearing points of pressure over the anterior process of the calcaneus and the dorsolateral surface of the cuboideus, and a smaller multilocular bursa covering the neck of the astragalus and the lower portion of the external malleolus.

The anterior annular ligament was thick and strong. The external annular ligament consisted of a firm, broad band covering the peroneal tendons as they passed to the posterior surface of the calcaneus. The internal annular ligament was well marked but very short. The muscles in the leg and foot were atrophied and had the following relations : The tibialis anticus passed over the lower third of the tibia, from without inward, to the inner surface of the internal malleolus, whence it descended to its insertion. The extensor proprius pollicis passed through a separate compartment of the annular ligament over the inner surface of the internal malleolus, close to and in front of the tibialis anticus. It divided into two tendons which were inserted in the second phalanx.

The extensor longus digitorum passed over the outer third of the anterior surface of the tibia, in a groove bounded by two well marked prominences. It was covered by the annular ligament. It divided into four tendons, which passed obliquely inward over the heads of the first and second metatarsal bones to the four toes.

The peroneus tertius passed obliquely outward to the base of the fifth metatarsal bone over the cuneiform bones. The extensor brevis digitorum was scarcely observable except for its tendons.

The peroneus longus and brevis passed beneath a strong and narrow band of fascia, extending from the external malleolus to the calcaneus, and representing the external annular ligament, around the external and posterior surface of the calcaneus to their insertions. The former did not touch the cuboid bones. It passed above the level of the anterior processes of the calcaneus obliquely, directed from without downward and inward. The latter after passing over the anterior process of the calcaneus, descended upon the lateral border of the cuboid bone to its insertion.

The tibialis posticus passed to the inner side of the internal malleolus, its posterior border beneath it, and to its insertion.

The flexor longus digitorum passed to the outer side of the tibialis posticus, beneath the internal malleolus, to its insertion.

The flexor longus pollicis passed through a well marked groove in the tibia, external to the flexor longus digitorum, beneath the internal malleolus, and was lost in the muscles of the foot. The relations of the bones of the foot to one another and to those of the leg, were in general as follows:

The position of the calcaneus was one of marked plantar flexion. Its long axis formed with the articular surface of the tibia an angle of 80°, which amounts to 41° of forced plantar flexion of the foot. Further, the bone was supinated 50°, and adducted about 10°. Besides the articular surface for the astragalus, there was upon this surface, just in front of the tuberosity, nearthroses for both the tibia and fibula. That for the former was continuous with the external third of the articular surface for the astragalus, and was situated upon the superior (really internal) surface. That for the latter was situated upon the external (really superior) surface.

The anterior process of the calcaneus was large and prominent. Situated upon the internal surface of the neck of the calcaneus was the articulation for the cuboid.

The astragalus was situated in the angle formed by the bones of the leg and the calcaneus. Its superior surface articulated with the tibia, and was external so far posterior as to be continuous with the inferior articular surface in its outer half of the same surface by a 5 mm. strip of nonarticular bone.

The shape of the body was triangular with the apex posterior, and with the surface looking upward and the inferior surface looking downward and backward. The neck of the astragalus formed with the body an angle of 45° for the inward displacement, and of 90° for the downward displacement. Upon the internal surface of the neck is an oval facet for articulation with the scaphoid bone, the long axis of which is placed at an angle of 45° to the long axis of the neck.

The scaphoid bone articulates with the neck of the astragalus only in the external portion of its superior surface; with the internal portion of the same surface it is nonarticular and is bent sharply upon the outer segment. The long (that is, transverse) axis of the scaphoid is nearly parallel with that of the neck of the astragalus. The bone presents two nearthroses. One for the anterior surface of the sustentaculum tali; the other for the internal malleolus at its anterior and inferior angle. The superior (anterior) surface was broad. The inferior was narrow. The external was narrower than the internal. There was no tuberosity present, nor did the bone articulate with the cuboid. The transverse axis formed with the transverse axis of the cuboid an angle of 90°.

The cuboid bone articulated with the anterior pro-cess (that is, neck) of the calcaneus upon its internal surface, in such a manner that the weight of the body was in part sustained by the dorsolateral surface. There was no sulcus for the peroneus longus tendon, nor was the tuberosity present. The three cuneiform bones articulated with the scaphoid and cuboid in the angle formed by them. Their position was one of marked supination, adduction, and approximation of the inner and outer bones toward the plantar surface.

The tibia and fibula were rotated inward about this long axis with the external malleolus anterior to the internal. Upon the fibular malleolus were two facets, one for the astragalus and one (a nearthrosis) for the calcaneus. These were separated by a nonarticular surface. Upon the tibial malleolus were three articular facets, one for the astragalus, and one (a nearthrosis) for the scaphoid, and one (a nearthrosis) for the internal cuneiform.

For comparison with Hartley's adult case, we will review the findings of R. W. Parker, of London, England, in a patient of 18 months, who died of tuberculous meningitis.

The spinal cord was normal to the eye and on microscopic examination; as was also the popliteal nerve and its main divisions. Portions of each muscle of the leg were examined microscopically and found to be perfectly healthy. The anterior portion of the internal lateral ligament of the ankle-joint was firmly blended with the astragaloscaphoid ligament above, and the calcaneoscaphoid ligament below, and all were shortened. A bursa was found between the tip of the malleolus and the scaphoid bone. The upper articular surface of the astragalus extended backward as far as the .posterior surface of the lower articular surface. The neck of the astragalus was elongated and its obliquity amounted to 53°, as against 38° in the normal bone. The articular surface of the head was prolonged to the inner side. The internal malleolar facet was not recognized. The calcaneus was rotated inward beneath the astragalus and a considerable portion of the upper posterior facet was uncovered and articulated with the posterior border of the external malleolus. The plane of the cuboidal facet was directed unnaturally inward.

The symptoms of congenital club-foot are the evident deformity, the restricted range of motion, a certain amount of muscular atrophy from disuse, and callosities from weight-bearing in an unnatural position.

The diagnosis of congenital clubfoot presents no difficulties, providing the history can be relied upon, except as to the anatomic relations and the consequent choice of treatment. Paralytic cases and at times hysterical cases present much the same appearance as congenital cases, but the history generally clears up the diagnosis at once.

The prognosis depends upon the degree of deformity and upon the treatment. Without treatment congenital cases grow progressively worse. With sufficiently prolonged mechanical support and such operations as the gravity of the case demands all deformities may be made right and the patient enabled to walk upon the soles of his feet. Improved muscular strength and muscular development are the natural result of use when once the deformities have been corrected. Clubfoot has no effect upon the general health and does not endanger life.

The treatment of clubfoot is mechanical, or operative, or both mechanical and operative. The earliest record that we have is that of the treatment employed by Hippocrates and consisted of bandaging and the use of a leaden shoe.

The first mechanical device intended to have a positive corrective action of which we have record is the "shoe" of Scarpa, of Pavia, in 1803, from which all

later shoes and braces for the treatment of this deformity
are hardly more than modifications. The aim of all
bandaging and brace treatment is to maintain the foot
in a somewhat corrected position, and to gradually gain
something upon this position by the exercise of so
much of a corrective force as can be borne without
serious discomfort or injury to the soft parts. Braces,
bandaging, and the various retentive dressings are used

FIG. 246.—Hand stretching of equinovarus during infancy.

either alone or in conjunction with various operative
procedures.

The simplest and most frequently used of the retentive dressings is the plaster-of-paris bandage. Plaster-
of-paris was first used in these cases by Jules Guérin, of
Paris, in 1826. He braced the foot in the best position
possible and poured liquid plaster around it. The
plaster bandage is of much more recent date. It is

used as a retentive dressing after hand-stretching and wrenching, and after any of the cutting operations to be mentioned hereafter. Narrow bandages from two to three inches wide are the best. They may be applied directly over the well greased foot, or over the foot wrapped with cotton or with bandages made from sheet-wadding. The most convenient covering, however, is a smoothly fitting stocking. The plaster bandage should be wrapped from within under the foot and outward so that at each turn the foot is drawn somewhat in the direction of correction of the deformity. When finished

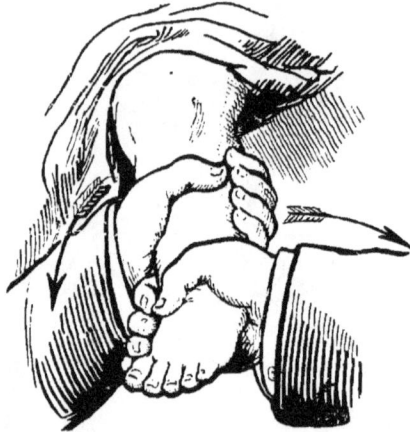

FIG. 247.—Hand-stretching of equinovarus in infancy.

the plaster splint should reach from the base of the toes to the garter line, and the foot must be held in the desired position until the plaster sets. The toes should be left exposed for an index that the limb has not been injuriously constricted. When used as a retentive dressing to hand-stretching or wrenching the plaster splint should be renewed every few days. When used after a cutting operation it is usually left on until the wound has healed and some more convenient retention device has been applied. In infants the plaster splint will be

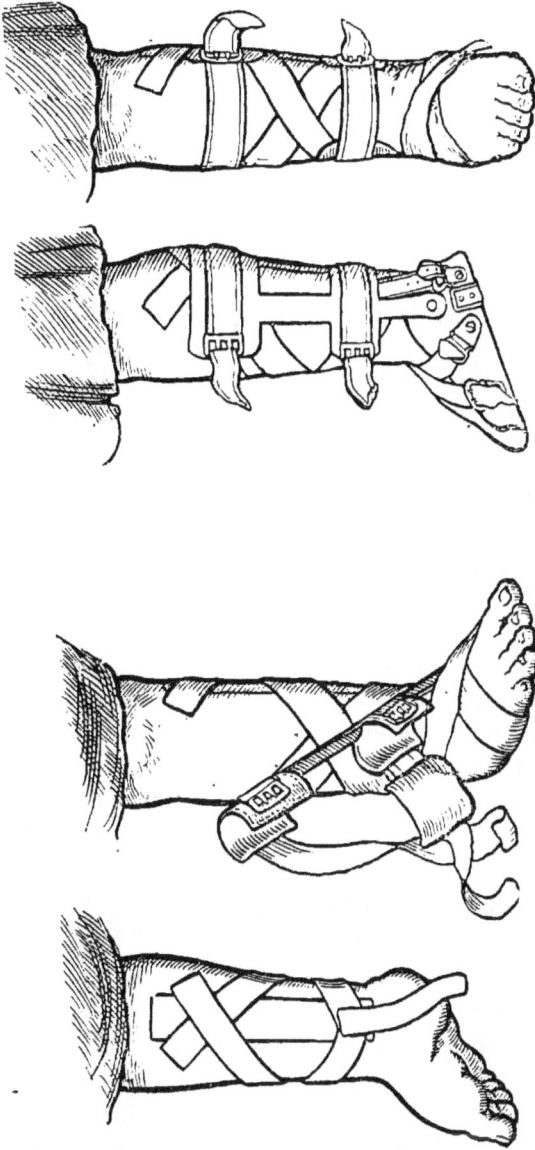

FIG. 248.—Continuous leverage treatment by the Taylor brace.

found a satisfactory dressing up to the time when the child attempts to walk, when a retention-splint having a flat sole and permitting of dorsal flexion at the ankle-joint will be required.

Barwell's dressing consists of a broad piece of adhesive plaster passed around the foot, from the dorsum around the inner margin of the front portion of the foot under the sole and up on the outer margin; this is connected by an elastic accumulator with a hook on the anterior and outer side of the shin at the garter line. The hook is usually made fast in its place by adhesive plaster. There are various modifications of this dressing, the more common being the use of two narrow strips of plaster instead of one passed around the foot and carried up the shin where they are made fast, and the whole covered by a roller-bandage.

Taylor's shoe consists of a foot-piece of sheet-steel made after a pattern of the sole of the foot, with a flange turned up against the inner margin. This flange reaches to the tip of the inner malleolus at the back and slopes forward a little below the level of the dorsum of the foot. The foot is held firmly in this foot-plate by various webbing straps and by adhesive plaster applied to the leg and passing down to a buckle at the upper and posterior angle of the flange of the foot-piece. The leverage for overcoming the varus is had by a leg-piece hinged to the foot-piece by a single rivet, and passing up the inner side of the leg to the garter line, where it terminates in a band encircling the leg. This band is of leather excepting that third which lies to the inner side of the leg, which is of sheet-steel. A similar band passes around the ankle. Just in front of the junction of the leg-piece with the foot-piece is placed a set-screw which limits the movement between the leg-piece and the foot-piece, permitting the foot to be dorsal flexed or the leg-piece to be carried forward, but not permitting the opposite movements. This brace is the most satisfactory of the continuous leverage braces, but it requires constant attention on the part of the surgeon to be at all useful.

Intermittent machine-stretching and continuous retention is best illustrated by Shaffer's shoes. In equinovarus the lateral deformity is treated first. Some years ago Shaffer used a shoe consisting of a sheet-steel foot-plate hinged obliquely at the ankle to a sheet-steel leg-plate, the correction force being a steel finger worked by a section of a toothed wheel and an endless screw. The shoe was bandaged to the inner side of the foot and the leg. This was modified by

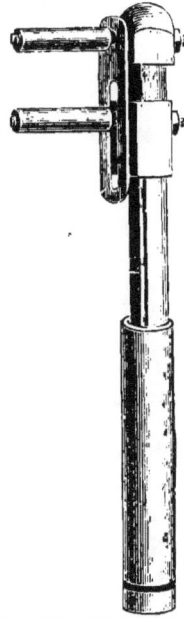

FIG. 249.—The Thomas club-foot wrench.

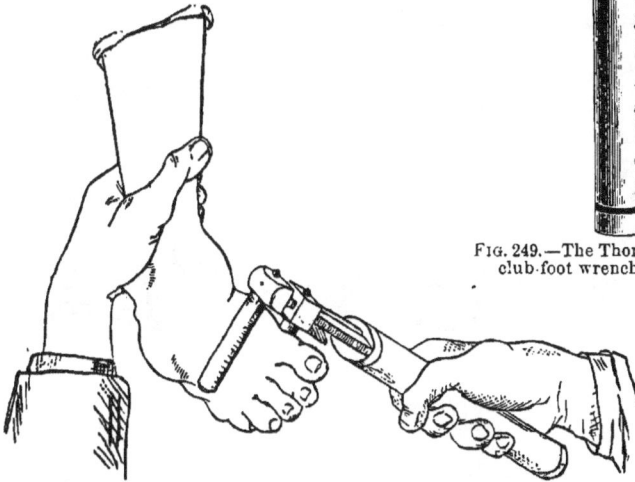

FIG. 250.—The Thomas clubfoot wrench applied to a case of equinovarus.

the writer, who placed the hinge at the outer side of the sole so that the inner margin of the foot was stretched instead of the outer margin being crowded together when the shoe was in use. Dr. Thomas L.

Stedman further improved the mechanism by substi-
tuting a rack-and-pinion for the worm-and-screw action.
These improvements are embodied in the shoe which
Dr. Shaffer now uses and which is applied to the outer
side of the leg and foot and is controlled by a ratchet-
and-pinion action.

When the lateral deformity has been corrected, the

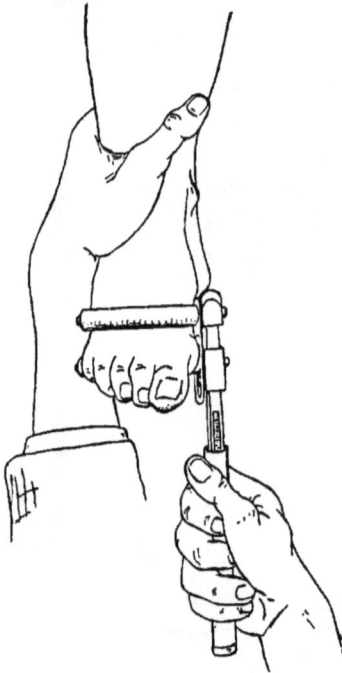

FIG. 251.—Twisting the foot with the
Thomas wrench.

FIG. 252.—Twisting and dorsal flexing
the foot with the Thomas wrench.

anteroposterior deformity is attacked by a shoe consisting of three parts and two actions. A steel band
passes around the back of the leg below the knee; from
this, two bars pass down, one on each lateral side of
the leg to the malleoli, to join the heel-cup by a worm-
and-screw mechanism. The foot is retained in the

heel-cup by a webbing-strap across the front of the ankle. The front half of the sole of the foot rests on a foot-plate which is joined to the bottom of the heel-cup by rack-and-pinion mechanism. A webbing-strap passes around the tip of the heel and passes down on each side of the foot to buckles on the bottom of the foot-plate.

In using either shoe the surgeon, daily, or oftener, when possible, stretches the foot toward the normal

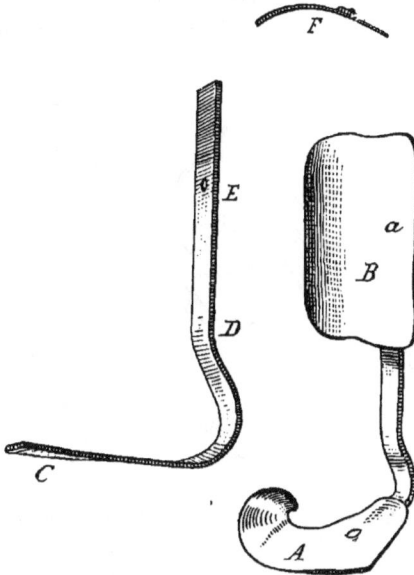

FIG. 253.—The Thomas clubfoot brace.

position up to the point of painful tolerance, holds it there for a short time, and relaxes the tension to the point of comfortable tolerance, where it is retained until again stretched.

Thomas's method of correcting the deformity by frequently repeated wrenchings with retention in the intervals was foreshadowed by Thomas Sheldrake, of London, in 1798. He says: "The essential operation to be performed in curing clubfoot is to produce such an extension of some of the ligaments, as, if it happened by accident, would produce a sprain." Thomas's method has not been generally understood, and consequently has been neither appreciated nor practised by any except a few of Thomas's pupils. The method consists of intermittent stretching, or wrenching, by a wrench and retention in simple iron splint. It is the method *par excellence* for the treatment of young children among the poor, where for any reason tenotomy may not be performed. It is applicable to all degrees

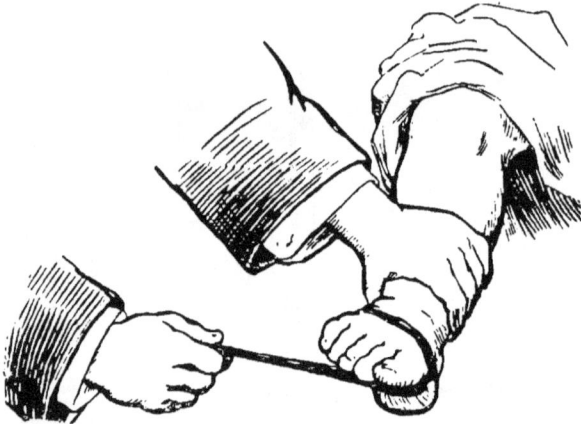

Fig. 254.—The first step in applying the Thomas clubfoot brace.

of deformity in young children; it gives as good results as can be obtained by any other method, or better; it more rapidly corrects the deformity than any other nonoperative method; and the cost is inconsiderable. But it is a cruel method, and one that cannot be employed in many private cases. Many parents prefer a longer course of treatment, or the risks attendant upon an operation, to subjecting their child repeatedly to the painful wrenching.

The wrench, Fig. 249, is made from a monkey-wrench

by sawing off the jaws of the wrench, boring a hole from the side through the fixed head-piece into which is set a strong pin and a like hole into the traveling head-piece into which is set a second pin. A slot must also be cut in the main stem of the wrench for the second pin to play through as the traveling head-piece moves up and down. A thin, slotted shield is placed at the base of the pins that the skin may not be pinched between the head-pieces as they approach each other. The pins should be straight and slightly bulbous at the ends to prevent their covers from slipping off. The pins may be snugly covered with thick, soft leather or with soft rubber.

The wrench is applied to the foot, as shown in Fig. 250; the foot is twisted and bent in the normal direction, as shown in Figs. 251 and 252. The twisting and bending is done forcibly and quickly, and the foot immediately released. Holding the foot too long in the bite of the wrench may result in a pressure-sore. The keynote of this treatment is the extent to which the stretching or spraining is carried. The wrenching should be carried far enough to temporarily destroy the resiliency of the soft parts; to such a degree that the foot is temporarily paralyzed and lies limp in the hand of the operator. It is then placed in the retention-brace in its best possible position and held there by adhesive bandages. After two or three or more days, depending upon the severity of the deformity and the severity of the wrenching, the resiliency of the soft parts begins to return; the foot is then subjected again to the wrenching procedure. In this way the treatment is continued until the deformity is fully corrected and until it shows no tendency to return. After this the foot is retained in the corrected position until all the parts have adapted themselves to their new relations; or, as Thomas used to say, "until the slack had been taken up" on the side of the convexity. This taking up of the slack is the second essential feature in this treatment.

Surgeons have always known that corrected clubfeet were prone to relapse unless retained in the corrected position for a certain time, but the reason for the relapse, and a positive rule for diagnosticating a cure, have not heretofore been made known. It has been assumed that the stretched, torn or cut parts on the

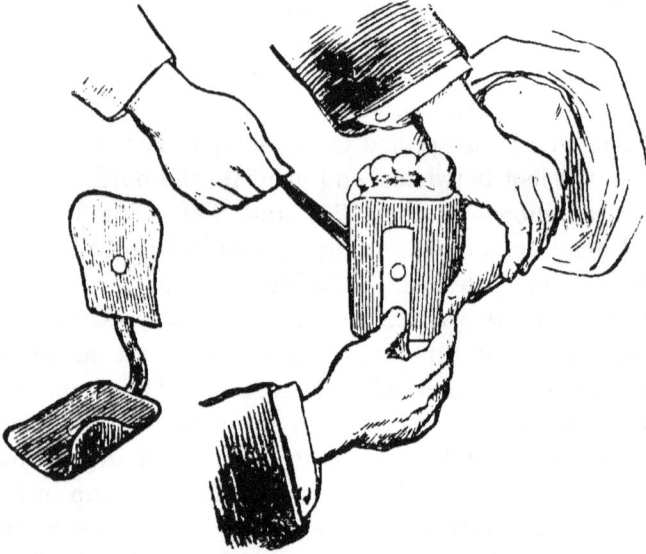

FIG. 255.—The second step in applying the Thomas clubfoot brace.

side of the concavity have recontracted, but no worthy explanation has been advanced for this recontraction, and some have assumed that it was owing to some inherent vice of these parts. The fact of the matter is, that the deformity returns from the same cause as a like deformity develops in noncongenital cases, namely, from a lack of muscular balance.

In the noncongenital cases of equinovarus, the fault is a paralysis, and lies in the abductors and dorsal flexors of the foot; in the congenital cases the fault, to which the relapse is due, lies in these same muscles; they are weak from disuse, and at a disadvantage from overlengthening. When the foot has been continuously held in a corrected or overcorrected position for a sufficiently prolonged period, structural shortening takes place in these elongated muscles, and in the ligamentous and other soft parts on the convexity of the deformity. When, then, the foot has been retained in the desired position so long that it cannot by manipulation be carried into the old deformed position any more readily than into the opposite defority, then, and not till then, can the treatment be discontinued, the patient discharged cured, and the certainty of no relapse prognosticated.

Tenotomy.—In so far as we know, the first section of a tendon for the correction of clubfoot was an open division of the tendo-Achillis by Lorenz at the suggestion of Thilenius, of Frankfort, in 1784. The first subcutaneous tenotomy is claimed to have been done by Mark Anthony Petit, in England, in 1799. But it appears that the

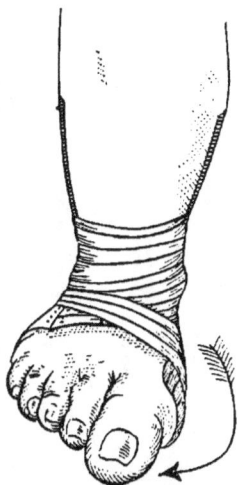

FIG. 256.—The Thomas club-foot brace applied.

operation was not repeated, for more than 30 years later William J. Little could find no one in England to undertake the operation. On the continent, Michaelis operated in 1811, Sartorius in 1812, Delpech, of Montpellier, France, did subcutaneous tenotomy of the tendo-Achillis in 1816. But to Stromeyer, of Hanover, more than to any other, do we owe the development of the operation in 1831. To Stromeyer went Dr. William J. Little, of London, for the cure of his own foot when he could find no one in England willing to do the new operation. Little introduced the operation into London, in 1836.

In America the first subcutaneous tenotomy for the cure of clubfoot was made by Dr. David L. Rodgers, of New York, in 1834, and the next by Dr. James H. Dickson, of North Carolina, in 1835. Dr. William Detmold, of New York, operated in 1840, and did more to popularize the operation in this country than any other surgeon. He was followed by David Prince, who wrote in 1866, Louis Bauer, 1868, and Lewis A. Sayre, in 1875. The comparatively recent date at which this operation was generally accepted may be realized by an editorial note in Dr. Townsend's translation of Velpeau's "Surgery," edited by Valentine Mott, in 1847. After reviewing the operation the editor says: "Notwithstanding the facts, the question of tenotomy still remains undecided."

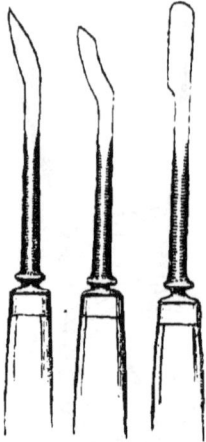

FIG. 257.—The tenotomes used by Mr. Robert Jones.

The operation as performed by most surgeons in England and America at the present day is practically the same as that recommended by Stromeyer in 1831. It consists in correcting the lateral deformity (the varus) in so far as it is possible before attempting the division of the Achilles tendon. The tendon is then divided with a tenotome, a small knife with a short and narrow blade with a long shank. The tenotome is inserted flat beneath the tendon, its edge is then turned towards the tendon which it is made to divide by a sawing motion. Care is taken not to separate the cut ends of the tendon until soft union has taken place at the end of from

4 to 7 days. Then by manipulation and retention it is attempted to stretch the plastic material which unites the cut ends of the tendon until the desired position has been gained, the fear being that the cut ends of the tendon will not unite if they become too widely separated. As early, however, as 1838 Scoutetten advised the immediate correction of the deformity after tenotomy, and this for many years has been practised by Sayre, in New York, and R. W. Parker, in London. That distant separation of the cut ends of the tendon does not of itself cause nonunion we are sure, having obtained solid union where the separation, in a young woman, was 1¾ inches.

Another fault in the old operation was the attempt at correction of the lateral deformity before the anteroposterior defect had been remedied. As pointed out in the remarks on the anatomy of clubfoot the body of the astragulus is often wedge shaped, with the apex of the wedge posteriorly, and there is, consequently, a short posterior ligament at the ankle-joint. Unless this ligament is cut, a somewhat difficult task, or torn before the tarsus itself is weakened by an operation, it will be found practically impossible to fully flex the foot; and unless full flexion well past the right angle be possible, relapse of the varus is sure to occur, for only by adducting the foot at the tarsal joint is the patient able to bring the heel to the ground.

The procedure, then, which we would advise, in the ordinary case of congenital equinovarus, is as follows : The foot is made clean; the patient is fully anesthetized, and turned on his face; an assistant grasps the leg below the knee and dorsal-flexes the foot; the skin at the back of the ankle is drawn somewhat to one side and a puncture of the skin is made with the point of the tenotome at the side of the tendon about an inch above its insertion; the puncture is carried across between the tendon and the skin; the knife is withdrawn and replaced by a blunt-pointed tenotome; this is then turned with its edge to the tendon and the tendon cut, mainly by the pressure exerted on the back of the knife with the thumb of the other hand. Of course the knife can be entered below the tendon and the cut made towards the skin, but in doing this it is easy to puncture the tendon when intending to pass beneath it, in which case a second puncture will need to be made, and in the ankle of a fat baby it is possible to mistake the fibula for the tendon and find that a section is not possible. In skilled hands the tendon may be cut with the sharp-pointed tenotome that is used to make the puncture, but it is hardly as safe as using the blunt instrument. As soon as the assistant feels the snap of the divided tendon he should instantly relax the flexion. When the cavity left by the incision has filled with blood the operator attempts full flexion of the foot. In a certain number of cases this will not be found possible, and the choice then is between a section or a forcible rupture of the posterior ligament of the ankle-joint. Section of this ligament is accomplished by a spear-shaped tenotome passed through the middle of the tendo-Achillis, then turned half around and swept from side to side. We prefer, however, to rupture this ligament. The knee is flexed to a right angle and rests on the table, the leg being vertical, then with the hand grasping the foot the weight of the operator can be thrown upon it and the ligament readily ruptured. Unless full flexion at the ankle-joint be rendered possible before the anterior of the foot is subjected to operation it will be found very difficult of accomplishment.

The next step is the division of all tight bands on the inner aspect of the foot in the neighborhood of the inner malleolus. The sharp pointed tenotome is inserted in front of the inner malleolus at about the point where the anterior tibial tendon passes to the inner side of the foot and passed around under the malleolus closely hugging that bone; the edge of the knife is then turned towards the shortened deltoid ligament which is divided as the knife is withdrawn. Through this same opening the tendon of the posterior tibial and that of the anterior tibial and the astragalo-scaphoid capsule may be divided. The foot is then straightened and rotated outward and everted by forcible manipulation or the use of the Thomas wrench or the osteoclast. In a few instances the plantar fascia and the adductor of the great toe will require division through a puncture at the side of the sole in front of the lesser tuberosity of the os calcis. There is only one rule as to what should be divided, namely, all tight bands that can be distinguished by the touch; after that every thing else that resists a full or even overcorrection of the deformity should be torn by the hand or the wrench or the osteoclast. We usually seal the wounds with collodion and a pledget of cotton. In the tarsal region troublesome bleeding may be safely checked by a pressure compress, but no pressure should be placed over the divided Achilles tendon. Pressure diminishes the quantity of plastic effusion and may be the cause of week union. A few layers of aseptic gauze are then placed over the wounds, the foot and leg enveloped in a wadding bandage and the whole put up in a plaster bandage from the toes to the knee, or above. The toes should be left exposed for an index of the circulation. The plaster is changed on the third or fourth day, when the wounds will usually be found healed. In infants the plaster-dressing, changed about every 2 weeks, is continued until the child begins to walk, when it is replaced by a retention brace. In older children the brace is applied as soon as the foot readily remains in the desired position. The retention brace which we would recommend is either the Thomas clubfoot shoe, already described, or a brace consisting of a steel foot-plate with a flange turned up at the inner side, to which is attached a side-bar running up the inner side of the leg to the garter line and thence half way around the leg, the full circumference being completed by a strap and buckle. The side-bar and foot-plate are joined by a single strong rivet, set loosely so as to allow motion, which motion is limited to a right angle in extension by a stop set in the flange of the foot-plate. A narrow flange is usually turned up at the outer side of the heel, from which a strap of webbing passes across the back above the tip of the heel, and the foot is held securely in place by another webbing strap passing from underneath the heel around the outer side of the foot and across the instep to a buckle set on the inner flange of the foot-plate. This brace is worn inside the stocking to more securely grasp the foot. The brace shou'd be worn until the foot shows no tendency to return to the deformed position, at least for 18 months, or even for a much longer period.

Phelps' operation consists in making an open incision commencing in front of the inner malleolus and extending one-third the distance across the sole of the foot, carried down to the neck of the astragalus, on its inner side. Through this wound the adductor pollicis, tibialis posticus, the plantar fascia, the flexor brevis, the long flexor tendons of the toe, and the deltoid ligament, all its branches if necessary, can be cut. This is done after subcutaneous tenotomy of the tendo-Achillis has been performed. Great force is then used by the machine shown in Fig. 259, to rupture the deep ligaments and supercorrect the foot.

Any case that can be corrected by the hand, or by subcutaneous tenotomy, should not be subjected to

Phelps' operation ; and when the operation as described fails to easily supercorrect the foot, a linear osteotomy should be made through the neck of the astragalus. This failing, the removal of the cuboid and scaphoid is indicated. And, as a last resort, Pirogoff's amputation should be resorted to.

The wounds of Phelps' operation are sutured, or not, as may be possible, and dressed without drainage with the aim to obtain blood-clot organization. The feet are dressed in supercorrected position in plaster-of-paris.

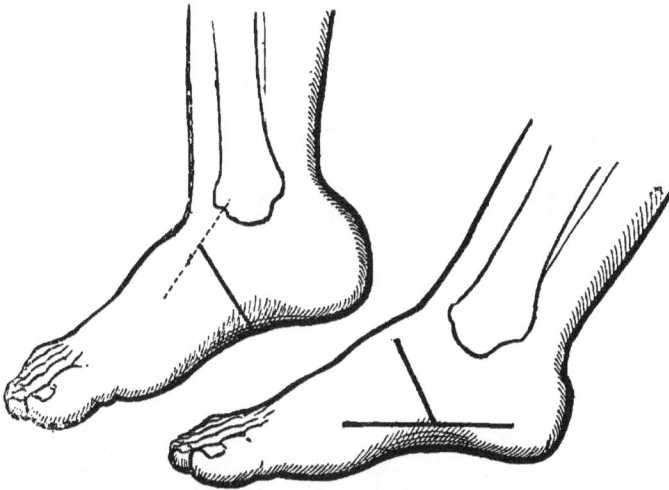

FIG. 258.—The heavy line in the figure on the left shows line of incision in Phelps' operation. The heavy lines in the figure on the right show lines of incision in author's operation.

Phelps first operated in 1879, and to him more than to any other is due the credit of leading the way to perfect results in inveterate cases of clubfoot.

Ridlon's operation is a modification of Phelps' operation, designed to avoid a tender scar of the sole of the foot. In Phelps' earlier operations, the incision was carried two-thirds across the sole of the foot; at present it is carried but one-third that distance, and the objec-

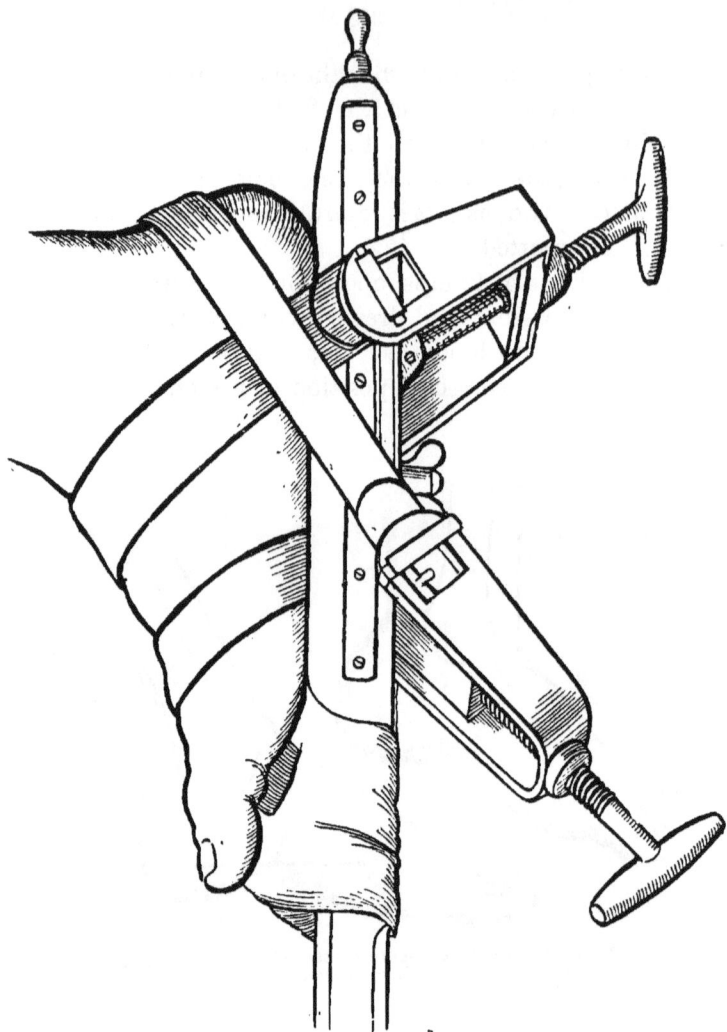

tion to a tender scar holds less well than before the modified operation was devised.

This operation consists of an incision commencing on the dorsum of the foot just in front of the inner malleolus at the point where the tendon of the extensor longus digitorum muscle crosses to the inner side of the foot; from here it is carried directly toward the sole, to meet, near its middle, a second incision made parallel with the sole from near the inner tuberosity of the os calcis to the middle of the first metatarsal bone or beyond. The plane of the first incision leads directly downward to the bones; the plane of the second incision slopes upward and outward to reach the bones at their nearest border, thence it is carried underneath the bones, closely hugging them. As these incisions are made, an assistant constantly keeps the parts on the stretch in the direction of correction of the deformity and each part as it appears to resist the correcting influence is divided. When any considerable degree of equinus is present the incision passes beneath the anterior half of the inner malleolus, and in severe cases beneath the whole of it. The deltoid ligament, and the tendons of the tibialis anticus, extensor proprius hallucis, extensor longus digitorum, extensor brevis digitorum, tibialis posticus, flexor longus digitorum and flexor longus hallucis are readily reached. From the second incision the abductor hallucis, the plantar fascia, and any other resisting structures may be divided. By carrying the second incision in the plane directed it will almost always be possible to avoid dividing the internal plantar artery and the nerve. The division of all tight bands having been made, the foot is overcorrected by the hand, or by the wrench. As in Phelps' method, we believe it better to divide the tendo-Achillis and correct the equinus, rupturing, if necessary, the posterior ligament of the ankle-joint before commencing the operation for

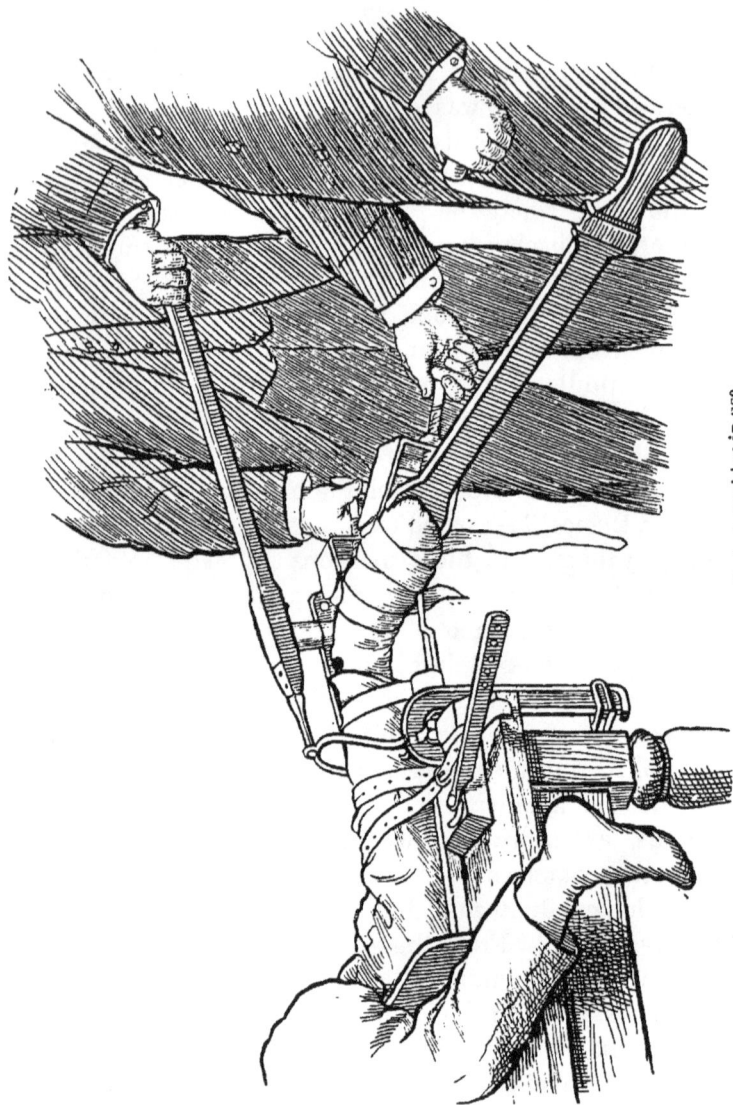

FIG. 260.—Phelps' machine in use.

correcting the deformity of the anterior portion of the foot. The dressings are the same as in Phelps' operation, namely, suturing in so far as it is possible without putting the parts at too great tension, and covering the wound with Lister protective and aseptic dressings, with the expectation that healing by blood-clot organization will take place. Over all is applied a plaster-of-paris dressing with the foot in the supercorrected position. In inveterate cases where the foot cannot be easily held in the supercorrected position, operations on the bones are indicated as pointed out under Phelps' operation, or Grattan's osteoclast may be used to crush the foot into the desired shape. Recently we have made use of this maneuver subsequent to subcutaneus tenotomy in preference to a bone-cutting operation.

The principles first clearly emphasized we believe by Phelps, must never be lost sight of: First, correct all cases possible by hand; second, correct all possible by subcutaneous incision; third, of the remaining cases, divide by an open incision all resisting soft parts, and with the hand or some mechanical device supercorrect the deformity; fourth, when this cannot be readily done, do a linear osteotomy of the neck of the astragalus, followed, if necessary, by a cuneiform osteotomy of the os calcis; if this is not sufficient, enucleate the astragalus, or remove any bone or portion of bone which blocks the way; and if this is not sufficient, amputate.

Linear osteotomy of the neck of the astragalus is demanded in inveterate cases in older children and adults where the inward twist of the head of the bone is so extreme that supercorrection cannot be made after free section of all the soft parts on the inner side of the foot. An incision is made on the outer surface of the dorsum of the foot over the neck of the astragalus and the bone is divided by an osteotome. The foot is placed in the desired position, the wound is closed without drainage and dressed in the usual way. In a few cases linear osteotomy of the astragalus will not be found sufficient to permit of overcorrection of the deformity. In such cases the incision is extended downward, the anterior portion of the outer surface of the os calcis is exposed, and a wedgeshaped piece removed from that bone just back of its anterior articular surface.

In a certain number of cases, relapse takes place because of the obliquity of this anterior articulating surface of the os calcis, even when the deformity as a whole has been readily corrected either by tenotomy or stretching. In such cases cuneiform osteotomy of the os calcis is indicated to prevent relapse.

In a few cases the astragalus is partially displaced forward so that the posterior portion of the lower articular surface of the tibia articulates with the os calcis, and the tibial articular surface of the astragalus is so changed in shape and direction that in all except very young infants restoration of full function in the direction of flexion is not possible except after the removal of that bone. When the head of the bone has been exposed by a curved incision over it, and separated from its attachments, it is grasped by a bone forceps and the rest of the ligamentous attachments can be most readily separated by the use of a grooved chisel. The wound should be dressed without drainage and with the foot in the supercorrected position. If the deformity cannot be sufficiently corrected by some one or all of the means indicated, so that a comely and useful foot results, amputation should be advised.

Excision of a large wedge from the outer side of the foot without division of the soft parts internally, and without regard to what bones or portion of the bones are included in the wedge, is no longer generally performed.

CONGENITAL DISLOCATION OF THE HIP.

Congenital dislocation, or displacement, at the hip is not of very frequent occurrence, being less frequently seen than congenital clubfoot. It was fairly well understood in France in the early part of the century, and some few cases were treated successfully. It was not recognized in England until the late Dr. Carnochan, of New York, pointed it out to the London surgeons in 1844. Dr. Carnochan's book, published in New York in 1850, was the first work on the subject in English.

The anatomic defect is found in the head and neck of the femur, in the acetabulum, and in the ligaments and muscles of the joint. In some young cases the changes are few and slight in character, except that the femoral head is displaced from the acetabulum. In other cases the changes are more marked. The head of the femur is often smaller than normal and imperfectly rounded in its shape. The femoral neck is shorter, and at times smaller than normal, and is joined to the shaft at less than the usual obtuse angle. At times it is a right angle and at times less than a right angle. In some old cases the head and neck have been found to have nearly disappeared. The acetabulum is usually smaller than normal, though it may be relatively as large as the head of the femur. In a considerable number of cases it is more or less triangular in shape, and it is always shallow and partly filled with fat and fibrous tissue. The ligamentum teres is stretched and smaller, or ruptured, or has entirely disappeared. The cotyloid ring is contracted so that it may not be possible for the femoral head to enter through it into the acetabulum. The capsular ligament is stretched in proportion to the extent of the displacement, it is more or less constricted, and may even be closed at the point of constriction in old cases. In a few old cases it has been found that the head of the bone had escaped from the capsule. There is structural shortening or lengthening in any of the muscles or all of the muscles that influence the movements at the hip. In some instances the shortening appears to be chiefly found in the short muscles, excepting the gluteal and the ilio-psoas, and in other instances it appears to be the long muscles that are chiefly at fault.

The displacement is almost always upwards and backwards, or upwards and outwards on the dorsum of the ilium in the direction which the femur usually takes in traumatic dislocation. Indeed dislocation in no other direction was recognized, except in conjunction with monstrosities, until one of the authors (J. R.) reported an upward dislocation in 1889. Since then one other case has been reported by Dr. Ridlon, two by Dr. A. M. Phelps, and one by Dr. DeForest Willard. These were all dislocations upward and somewhat forward.

The cause of congenital dislocation of the hip is not known. Carnochan believed that it was due to muscular retraction arising from irritation of the ganglionic centers of the cord. The other causes to which it has been attributed are morbid retraction of a portion of the muscular tissues resulting from a defect in or absence of some portion of the nervous centers, arrested development, aberration of the nutritive forces, effusion into the joint cavity, relaxation of the muscular and ligamentous structures, intrauterine pressure exerted upon the levers presented by the long bones, external violence while in utero, traumatism occurring at birth, and heredity. Dupuytren, quoted by Carnochan, records the case of Marguerite Grandas, of Nantes, who had bilateral congenital dislocation. Her two maternal aunts had bilateral congenital dislocation; also one paternal aunt had unilateral dislocation, and another paternal aunt was the mother of a daughter who had unilateral congenital dislocation. Marguerite bore a daughter who had unilateral dislocation. This daughter married a healthy man whose father had bilateral dislocation, and bore four children; two of these had unilateral dislocations.

Personally we have seen few cases in which the defect was also found in the mother or had been handed down to the children. It is more often found in girls than in boys, more than 80% being girls, and in this it differs from other hereditary defects which are more frequently found in boys. Injury at birth has been charged with the causation of a certain number of cases where the defect is unilateral, and where the bones, femoral head and acetabulum, have been found to nearly approach the natural size and shape; but in these it must be remembered that an easy labor is quite as often recorded as a difficult one. As above indicated, the dislocation may be of both hips or of only one. In our experience it is more frequently unilateral.

The symptoms are not sufficiently pronounced to attract attention until the child begins to walk, or until it has walked for some months. In unilateral cases the symptom that then attracts attention is a limp; in bilateral cases it is a waddling gait, a sway-back and a prominent abdomen. The limp, or the waddle which is simply a double limp, is due to the shortening and the insecurity of the femoral head for weight-bearing when displaced from its bony socket and resting on the buttock muscles. The leg is shortened from $\frac{1}{2}$ inch in infancy to $1\frac{3}{4}$ at 7 or 8 years of age and $2\frac{1}{2}$ to 3 inches in adult life. There will be found an upward displacement of the greater trochanter equal to the amount of shortening. If a cord be drawn from the anterior

superior spine of the ilium across the hip to the tuberosity of the ischium it will be found in the normal to
pass across the tip of the greater trochanter ; in the

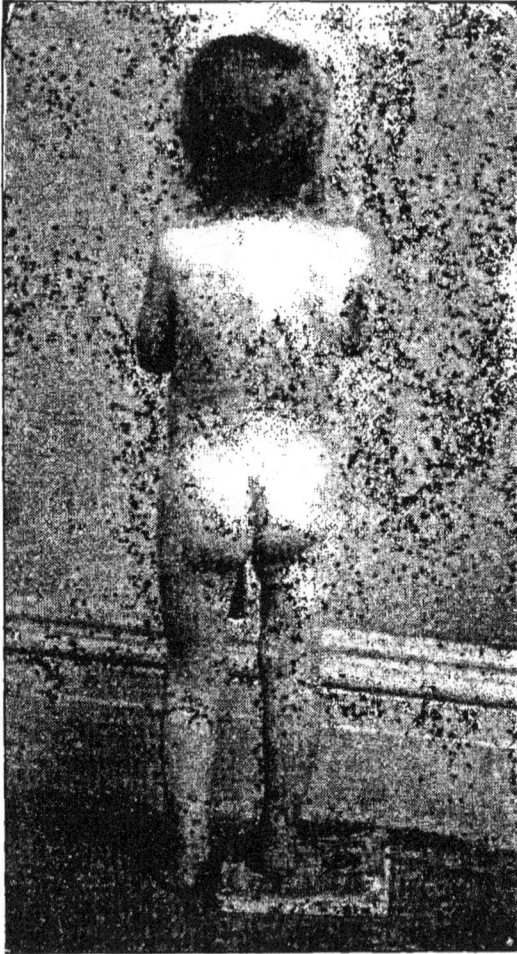

FIG. 261.—Congenital dislocation of the right hip.

congenitally dislocated hip it will pass below this point
by as much as the limb is shortened. The limb is

smaller and flabbier than the sound limb. The upper
portion of the buttock is prominent; the lower portion is
flat; the hip is broadened laterally; the perineum is
broadened, noticeably in bilateral cases; the pelvis is
tilted forwards, giving lumbar lordosis and a prominent
abdomen, more marked in bilateral than in unilateral
cases and increasing with age. All of these symptoms
are more noticeable when the patients stand than when
they lie. In cases of some standing a certain amount
of flexion deformity is found by the Thomas flexion
test. Movement at the hip in flexion, adduction and
inward rotation are as free and often are freer than
normal, while adduction, outward rotation and extension are restricted to some extent. In a word, any
motion that throws the femoral head against the dorsum
of the ilium is restricted, and any motion that throws
it away from the ilium is free. In standing, few patients
show any rotary deformity of the limbs, although most
writers note an outward rotation despite the fact that
anatomically it appears to us that there must be an
inward rotation unless there be twisting of the femoral
neck. There is rarely any complaint of pain or disability, except fatigue on long standing or walking, in
the case of children ; on the other hand, in old cases
there is often great disability, and patients at times take
to crutches.

The differential diagnosis from traumatic dislocation
can not be made except by the history of the case.
From. pathologic dislocations they are distinguished
by the history and by the freedom of motion in some
direction; in pathologic dislocations the pathologic
process which leads up to the dislocation restricts the
motion at the joint to some extent in all directions.
From infantile paralysis by the absence of flaccidity
in the distal muscles. Doubless it is possible for infantile paralysis to affect the buttock muscles alone, leav-

ing the distal muscles untouched, but we are not aware
that such a case has been reported. From congenital
shortening of the whole limb by the fact that the shortening is above the greater trochanter, while in congenital
shortening the whole limb is uniformly arrested in its

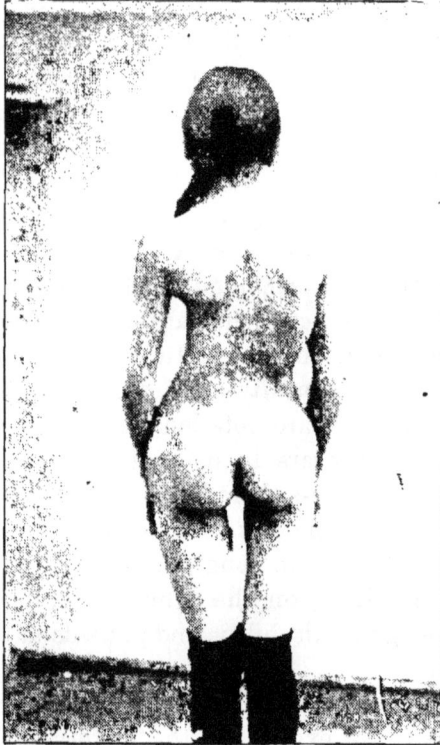

FIG. 262.—Congenital dislocation of both hips (with lateral curvature of the
spine); showing broadening of the hips and widening of the space between
the thighs.

growth. From fracture of the neck of the femur by
the absence of a history of injury and by the presence
of the head of the bone outside the acetabulum.
Fracture of the neck of the femur is not uncommon in
children, but we have not met with it in infancy. From

coxa vara by the presence of the head from the dorsum of the ilium, and by the history of the case that the child has always walked with a limp. We have not met with true coxa vara in children under five years of age. From hip disease there should be no difficulty in differentiating, for in hip disease motion at the hip is restricted to some extent in all directions.

As to the prognosis : It does not effect life. Without treatment these cases grow somewhat worse as time passes. Under treatment, some cases are cured, some cases are not benefited, and some cases are made worse.

The treatment has been of three kinds : First, mechanical, without any attempt at manipulative reduction of the dislocation; second, traction by hand or by machine to stretch the contracted soft parts followed by manipulative reduction, as in traumatic dislocation; and third, reduction in conjunction with a cutting operation.

The mechanical treatment has aimed to pull the leg down, and to prevent its being pushed up again. To pull the leg down an apparatus has been used either by weights in bed or by the conventional traction hip-splint. It has been protected from being pushed up again in walking either by some form of perineal crutch, or by a corset or girdle, made to press downward upon the trochanters, and by restricting the lordosis rendering the femurs less secure through the iliopsoas muscles. No permanent results have been had from this form of mechanical treatment.

Traction followed by manipulative reduction was first reported by Duval and Lafond in the case of a child of 9 years; but the permanency of the result is not reported. In 1835, Humbert and Jacquier reported successful cases. The reports were doubted. It was believed that they converted dorsal into ischiatic dislocations. Parvez, of Lyons, reported successful

cases. His work was investigated by a committee of
the Royal Academy of Medicine in 1836, and confirmed
by their report in 1838. This plan of treatment, however, fell into disuse despite a few favorable cases reported by Buckminster Brown, of Boston, William

FIG. 263.—Congenital dislocation of both hips; showing tilting forward of the
pelvis and lordosis of the lumbar spine.

Adams, of London, and Schede, of Hamburg, until
recently it has received a new impetus from the work
of Paci and of Lorenz. All cases are now subjected to
this treatment before being condemned to a cutting
operation.

Manipulative reduction, as now performed, consists
in pulling down the head of the bone and stretching
the soft parts either by the hands of assistants or by
mechanical means, the stretching process lasting several
minutes. One of the authors (J. R.) is accustomed to
fix the pelvis by passing a strong long towel between

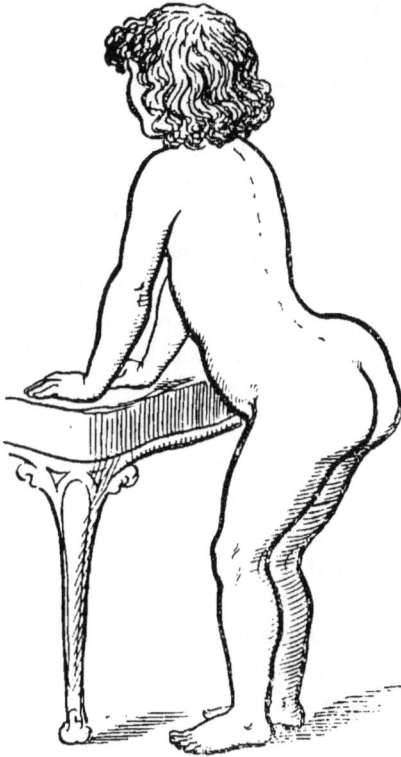

Fig. 264.—Congenital dislocation of both hips.

the legs of the patient, bringing up one end along the
groin and the other along the gluteo-femoral crease,
and wrapping the ends around the hands of the operator, who braces himself at the end of the table. Three
or four assistants grasp the leg and a tug-of-war ensues.

When the head of the femur has been pulled down as far as seems necessary, or as far as is possible, the pull is let up. Then the operator, standing on the side of the patient away from the joint, flexes the thigh to a right angle, winds his arm around the thigh from without, under, inward, and outward across the groin. Holding thus, he adducts, flexes, and lifts the head of the femur towards the acetabulum, and rotates it from side to side while he holds the pelvis down with the other hand. If this maneuver is successful the head will be felt to slip into the acetabulum with a distinct click, which at times can be heard; but oftentimes there is a false click, less distinct than that of the real replacement, which comes before it, and may delude the beginner. When the replacement has been effected the limb is carried into extreme abduction and held there while a plaster spica is applied from the knee to the scapula. The operator should always keep a hand upon the greater trochanter when carrying the thigh into abduction, for if the replacement has been a false one the head of the bone will be felt to slip upward again. If the replacement has been really accomplished it will remain in place fairly securely. Many operators direct that in putting on the plaster-dressing the limb be held abducted and rotated inward, but we have held it abducted and rotated outward, believing it a more secure position.

The first plaster dressing is kept on, if it remains firm, for about four weeks. The security of the head in the acetabulum may usually be readily determined by turning the patient on his face, placing the unoperated thigh in a like position to the one operated upon, and using the thumb and first fingers of both hands as calipers to try the relations between the tuberosity of the ischium and the greater trochanter on both sides.

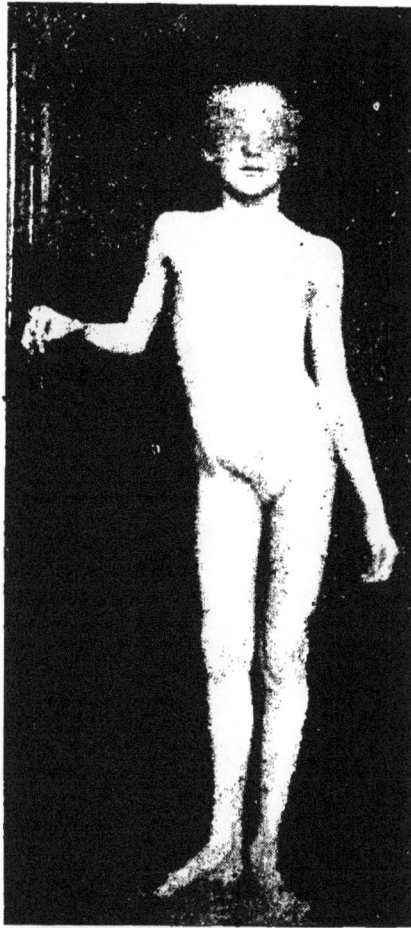

FIG. 265.—Congenital dislocation of the right hip.
The dislocation was upwards and slightly
forwards. This case was reported by Dr.
Ridlon, Nov. 16, 1888. In so far as we know
this is the first case of upward dislocation
reported. Since then one other case has been
reported by Dr. Ridlon, two by Dr. A. M.
Phelps, of New York, and one by Dr. De-
Forest Willard, of Philadelphia. This child
was 10 years old, and had had no treatment.
The shortening when standing was three
inches; when lying it was two and a half
inches.

A certain number of cases relapse. We have generally had a relapse in cases over five years of age, but have had only one relapse in younger children; that was a

Fig. 266.—Same patient shown in Fig. 265.

child of two years with unilateral dislocation whose mother had bilateral dislocation. We are accustomed to make a second or even a third replacement in case

343

of relapse, and if all relapse then there is nothing left to
do but a cutting operation. We have been accus-

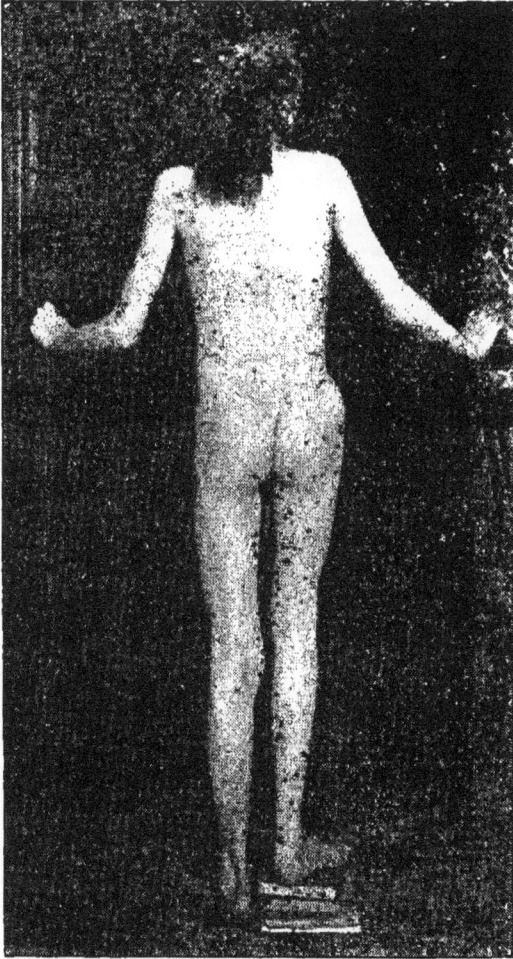

Fig. 267.—The same patient shown in Fig. 265, showing the
shortening when standing. The lines mark the iliac
crests and the greater trochanters.

tomed to keep patients in bed from four to eight
weeks before allowing them to walk, but we are

not sure that this is necessary. We have not made use of the high shoe on the opposite foot as has been recommended. We have continued the use of the plaster spica for 6 to 12 months. In some cases we have applied a light abduction splint after removal of the plaster. The splint consisted of a band about the chest, a band about the hips supplemented with perineal straps, and a band about the upper and another about the lower thigh, all bands connected by a bar at the side of the patient, hinged opposite the hip-joint for free anteroposterior motion and bent to the desired degree of abduction. The splint is more comfortable than the plaster spica, but it is less secure.

In no case have we had stiffness of the joint after the removal of the splint, or long after the removal of the plaster spica. We have not aimed, as Whitman has, to do as much harm as possible to the joint with the hope that adhesive inflammation would make the replacement more secure and with the desire to obtain a stiff and therefore a more stable joint. It appears to us a question whether a stiff hip is more to be desired than a freely movable one, though dislocated. In all cases there has been shortening of about a quarter of an inch.

The first operative work was in the way of excision of the joint; this did not prove a success. Guerin, in cases where it was found impossible to retain the head, divided the shortened muscles subcutaneously and scarified to provoke effusion of organizing material. His work was investigated by a commission from the Council General of the Civil Hospitals of Paris in 1843, and his cures were confirmed. This method, however, fell into disuse.

The popularization of the operative treatment for congenital dislocation of the hip is due to Hoffa, of Würzburg, who began this work about 10 years ago.

345

In 1891 we saw him operate, and the operation was substantially as follows : The patient was laid on his

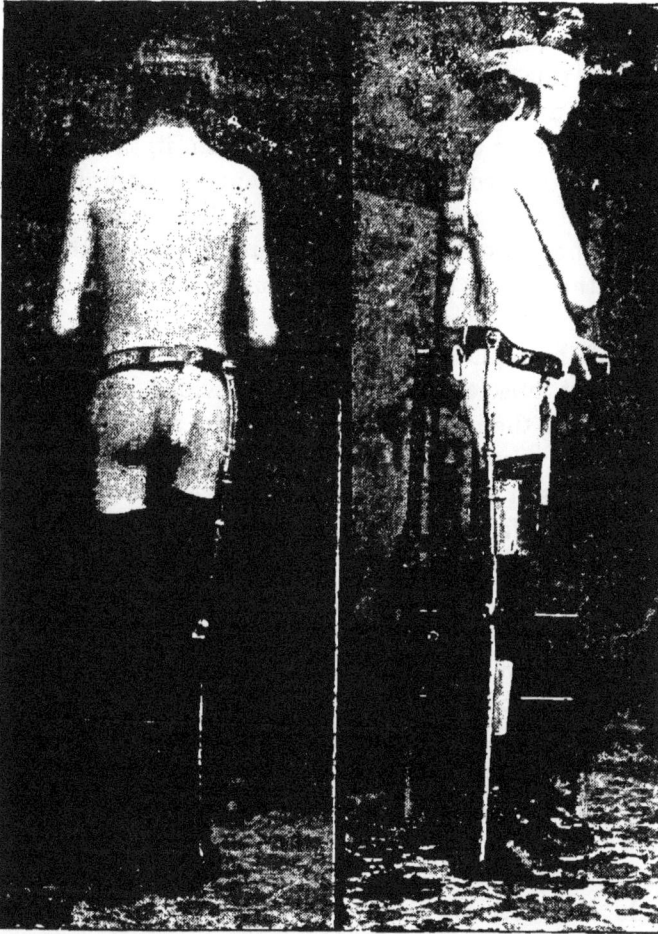

FIG. 268.—Showing the same patient shown in Figs. 265, 266 and 267, one year later after treatment in bed by traction. The lumbar lordosis has been overcome, as well as the flexion of the thigh, and the leg has been pulled down to the full length. A supporting walking splint is also shown.

side, the thigh flexed to a right angle, an incision following the line of the fibers of the gluteal muscle was made over and down to the joint, the capsule opened, the head of the femur turned out of the wound, all muscular attachments to the femur were separated sub-periosteally down to the lesser trochanter, the capsule was then followed with the finger and the acetabulum located, this was then enlarged with a sharp spoon or a gouge until it was large enough to easily receive the head and deep enough to readily retain it, then the leg was straightened. Hoffa then claimed that unless the short muscles were separated from the femur the re-placed head would be thrown out of the acetabulum, when the leg was straightened from the flexed position.

Lorenz, of Vienna, found that the obstacle to reten-tion of the replaced head was more often the long than the short muscles, and all these he divided, preliminary to opening the joint. He operated by an anterior in-stead of a posterior incision, and pared down the femoral head as well as enlarged the acetabulum. As a matter of fact it is sometimes one set of muscles and sometimes the other, and sometimes both that are at fault, and at other times it is not necessary to divide either, but better to use these shortened muscles as guys to securely retain the head in the acetabulum.

Bradford, of Boston, and Sherman, of San Francisco, have shown that the relapse from manipulative replace-ment and the difficulty in operative replacement lies in the contracted cotyloid ring and the constricted capsule, and have been successful when they limited their opera-tive procedure to opening the joint, slitting up the cap-sule and cotyloid ring, replacing the head in the acetabulum with no folds of ligament or capsule between it and the cavity of the joint, and then sewing up the capsule around the neck of the femur, and fixing the limb in the abducted position as in manipulative re-placement.

The Hoffa operation and all of its modifications that

call for the removal of bone either from the acetabulum or from the head of the femur are very serious operations, the mortality is considerable, and ankylosed joints, or worse, frequently result. The Bradford-Sherman operation, which removes no bone, promises to be a reasonably safe procedure when done by a cleanly surgeon. Infection of the wound in any of the cutting operations, however, is likely to result in caries and give a final result far worse than the original difficulty.

CONGENITAL DISLOCATION OF THE SHOULDER.

Congenital dislocation of the shoulder, properly so-called, is an exceedingly rare defect. Most of the cases that have in the past been called congenital dislocations at this joint have undoubtedly been traumatic dislocations occurring during the birth of the child. In true congenital dislocations the dislocation is said to be always subcoricoid or subacromial and the glenoid cavity is small and shallow, and the whole shoulder-joint is defective in its development. In the dislocations due to a traumatism inflicted at birth the structures of the joint are all approximately normal. The joint-capsule may or may not be torn and the glenoid cavity may be intact or a small portion of the rim of the cavity may be broken off and carried away with the displaced head of the humerus. These dislocations, in so far as we know, are always subspinous and the arm takes the same position and presents much the same restrictions to motion that will be found in connection with this dislocation when occurring later in life.

The arm is held close to the side, motion in all directions is restricted, muscular atrophy is not present to any extent, the limb is rotated inward so that the depression between the shoulder and the chest is deepened into a groove and the shoulder looked at from the front appears to be advanced; looked at from the back, however, it will readily be seen that the humerus is displaced backward. There is always a certain range of voluntary motion possible from the movement of the scapula.

The differential diagnosis is from tubercular disease of the shoulder, which presents the same restriction to motion, but which shows muscular atrophy and often

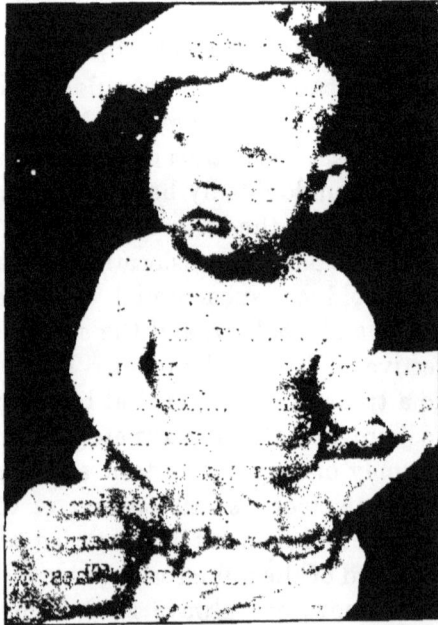

FIG. 269.—Congenital dislocation of right shoulder, probably due to injury at birth, showing characteristic position of adduction and inward rotation, and deepening of the crease between the shoulder and the chest.

swelling from thickening of the capsule, tenderness and pain, and does not show the rotary deformity and the groove between the shoulder and the chest; from infantile paralysis affecting the capsular muscles of the shoulder alone, and which presents the appearance of a downward displacement where there is no restriction

to passive motions while active movement is not possible; and from traumatic palsies which are usually
readily recognized from the history.

The treatment is replacement of the dislocated head
and retention for a period of many months. An attempt
at replacement during anesthesia should always be
made. In young cases it will often be successful; in
older cases it will often be a failure. Failure in attempted replacement and in retention after replacement
should always be supplemented by a cutting operation.
In truly congenital cases the cartilage of the head of
the humerus should be erased; in order that fibrous an-

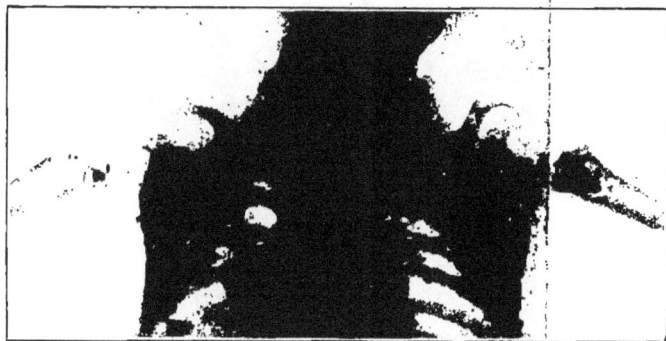

FIG. 270.—Röntgen picture of child in Fig. 269. Congenital dislocation of right shoulder.

kylosis may result; in traumatic cases it may or may
not be necessary to cut away some of the humeral head
to accomplish the reduction, then the loose capsule
should be gathered up and made firm, and the arm
put up somewhat abducted and with the elbow carried
far to the rear.

CONGENITAL RECURVATION OF THE KNEE.

Congenital recurvation of the knee has been described as congenital dislocation, but in those cases

that have come under our observation there has been no true dislocation. The legs are bent forward on the thighs to 45° or more, they can not be bent in the normal direction beyond the straight line, or thereabouts, and the patellae are absent or at least rudimentary bones. The condition is often associated with other congenital defects such as clubfoot and spina bifida.

FIG. 271.—A case of congenital recurvation of the knees. At birth the legs were flexed anteriorly on the thighs 45° and could be scarcely straightened. At the age of 21 months, when photographed, the knees could be flexed in the normal direction only as far as shown in the illustration. There was also present congenital knockknees, equinovarus, convergent squint, and spina bifida with incontinence of feces.

The treatment is by passive bendings and retention in the best possible position. Good results are to be expected, the patellae usually developing as the normal range of motion and voluntary use is gained.

Congenital Dislocation of the Patella.

Congenital displacement of the patella may be upward from an unduly long patellar ligament, or outward from a deficient outer condyle of the femur. The former is not as a rule a serious disability and will hardly prove a greater burden than the treatment. Outward displacement is a serious disability. It has been treated by various braces and trusses to retain the patella in its groove, but none are really satisfactory. A permanent cure can usually be readily effected by hammering the outer condyloid ridge with a rubber or a wooden mallet once a week until from periosteal irritation a ridge of sufficient height has been developed.

Congenital Constriction Bands.

Congenital constriction bands are rarely seen. We do not know what causes them, but they are met with in conjunction with other congenital deformities. They appear like narrow cicatricial bands closely hugging the bone and tend-

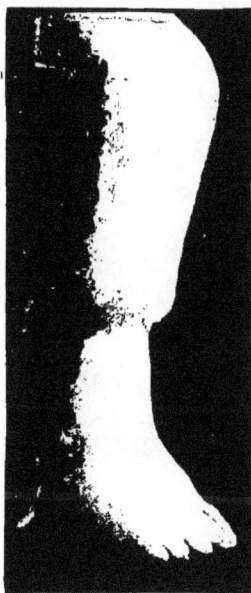

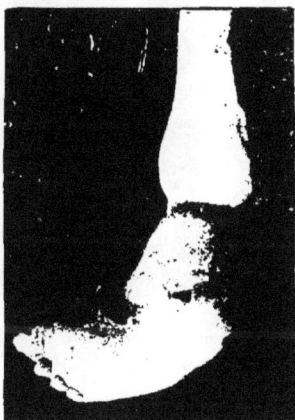

Fig. 272.   Fig. 273.

Fig. 272 shows the outer aspect and Fig. 273 the inner aspect of a congenital constriction band in a child 2½ years old. There was also present congenital equinovarus, webbed fingers, amputated fingers and constriction bands on one finger and two toes.

inous or ligamentous structures of an extremity. They
may be readily removed by dissection.

FIG. 274.—Same case as shown in Figs. 272 and 273 (same view as Fig 272)
after operation.

FIG. 275 shows same case as shown in Figs. 272 and 273 (same view as Fig. 273) after operation.

INDEX.

356

www.ingramcontent.com/pod-product-compliance
Lightning Source LLC
Chambersburg PA
CBHW021401210326
41599CB00011B/966